Lecture Notes in Mathematics

Edited by A. Dold an

547

Arunava Mukherjea
Nicolas A. Tserpes

Measures on Topological Semigroups:

Convolution Products and Random Walks

Springer-Verlag
Berlin · Heidelberg · New York 1976

Authors
Prof. Arunava Mukherjea
Prof. Nicolas A. Tserpes
University of South Florida
Department of Mathematics
Tampa, Florida 33620/USA

AMS Subject Classifications (1970): 43A05, 60G50, 60J15

ISBN 3-540-07987-4 Springer-Verlag Berlin · Heidelberg · New York
ISBN 0-387-07987-4 Springer-Verlag New York · Heidelberg · Berlin

Printed in Germany

Printing and binding: Beltz Offsetdruck, Hemsbach/Bergstr.

Preface

This monograph is an outgrowth of the lecture notes of a series of lectures given by the first author in the Indian Statistical Institute during the fall of 1973. These notes supplement in many ways the material presented in the book "Probabilities on Algebraic Structures" by Ulf Grenander and the material that appears in Chapters IV and V of the book "Markov Processes: Structure and Asymptotic Behavior" by M. Rosenblatt. Like most mathematicians who have worked in this area, we owe much to these two mathematicians. We also gratefully acknowledge a number of stimulating conversations with Prof. M. Rosenblatt when he was invited to speak in the Wayne State University Semigroup Symposium in 1968 and when he was invited to give a series of colloquium talks at the University of South Florida in early 1973.

Our primary objective in these notes is to provide the reader with a brief, but somewhat complete account of the theory of probability and measure on topological semigroups in the context of the following problems: (i) the characterization of the idempotent and r*-invariant probability measures on locally compact Hausdorff topological semigroups (ii) the limit behavior of the averaged and unaveraged sequence of convolution iterates of probability measures on different topological semigroups, and also on semigroups of stochastic matrices (iii) almost sure convergence of products of independent random variables taking values in a completely simple semigroup and (iv) the recurrence behavior of one-sided and two-sided random walks induced by a probability measure on a compact Hausdorff or locally compact Hausdorff completely simple topological semigroup. Thus our notes cover only certain aspects of probability theory on semigroups while leaving out many other interesting aspects such as the study of infinitely divisible probabilities on groups and semigroups and a discussion of the embedding problem for such measures. Other interesting subjects which we have not even touched include the study of potential theory for recurrent random walks initiated by Spitzer and later studied by Kesten, Ornstein, Port and Stone, Brunel and Revuz, and others. The main reason for these omissions is that these areas of study, while highly explored in the context of groups, have been somewhat overlooked in the general framework of topological semigroups.

We hope that the reader will find the results and the methods that are developed in these notes useful in many different contexts. We feel that these notes can be covered during a one-semester seminar meeting once a week for two hours in a typical American university.

We express our deep appreciation to Professor K. H. Hofmann of Tulane University and Professors A.T. Bharucha-Reid and T. C. Sun of Wayne State University. We have learned a great deal from them on semigroups, measures and probability, through occasional correspondence and actual collaboration .

The preparation of these notes is partially supported by the National Science Foundation.

A. Mukherjea

N. A. Tserpes

TABLE OF CONTENTS

CHAPTER I

MEASURES ON SEMIGROUPS

1. Introduction

It is well-known that every locally compact Hausdorff topological group admits of a left (as well as a right) invariant measure, which is regular in a certain sense. In recent years, the theory of semigroups and topological[++] semigroups has developed a great deal and consequently the study of measures in different contexts has been possible on these more general algebraic structures. It has been found that presence of certain measures (invariant measures, idempotent measures, etc.) on general topological semigroups impose on these semigroups certain definite structures. Also certain invariant measures and their structures play a definite role in the study of probability theory on semigroups; for instance, idempotent probability measures, which appear as limit distributions for the sequence of partial sums of an infinite sequence of independent, identically distributed random variables with values in a suitable topological semigroup, are closely associated with certain invariant measures.

The object of this study is to present the theory of idempotent and invariant measures on locally compact Hausdorff topological semigroups and different limit theorems involving probability measures and their convolutions.

[++] An algebraic semigroup S is topological if there is a topology on S such that the binary operation $(s,t) \to st$ is jointly continuous in s, t.

2. Preliminaries on Semigroups

Let S denote a semigroup, i.e. a non-empty set with a closed associative multiplication.

2.1 Definitions. A non-empty subset $I \subset S$ is called (i) a right ideal iff $IS \subset I$ (ii) a left ideal iff $SI \subset I$ (iii) an ideal iff $SI \subset I$, $IS \subset I$.

2.2 Example. [This example is quite useful and will be referred to often.] Let G be a group, X, Y be any two non-empty sets and ϕ be a function from $Y \times X$ into G. Then if we define multiplication in $X \times G \times Y$ by

$$(x_1, g_1, y_1)\ (x_2, g_2, y_2) = (x_1, g_1\ \phi(y_1, x_2)\ g_2,\ y_2),$$

$X \times G \times Y$ becomes a semigroup. This semigroup doesn't have any proper ideals. If $A \subset X$, then $A \times G \times Y$ is a right ideal. If A is a singleton, then it is a minimal right ideal. Similarly, if $B \subset Y$, $X \times G \times B$ is a left ideal; if B is a singleton, it is a minimal left ideal.

2.3 Definitions. (i) e is called an idempotent in S if $e = e^2 \in S$. An idempotent e in S is called (a) a right identity iff $se = s, \forall\, s \in S$; (b) a left identity iff $es = s, \forall\, s \in S$; (c) a left zero iff $es = e, \forall s \in S$; (d) a right zero iff $se = e,\ \forall\, s \in S$.
(e) an identity iff $se = es = s, \forall\, s \in S$ (f) a zero iff $es = se = e, \forall\, s \in S$.

(ii) S is called right simple iff S has no proper right ideals, i.e. if $xS = S\ \forall\ x \in S$. [Note that xS, $x \in S$, is a right ideal of S and for any right ideal I of S, $xS \subseteq I$ for $x \in I$.]

(iii) S is called left cancellative iff $xz = yz \Rightarrow x = y$, $\forall x,y,z \in S$.

(iv) S is called a <u>right group</u> iff S is right simple and left cancellative. [Note : S is a right group $\Longleftrightarrow$ S is right simple and contains an idempotent.

<u>Proof</u>. Suppose S is a right group. If $x \in S$, then $xS = S$, hence $\exists\, y \in S \ni x.y = x$ or $x.y^2 = x.y$. Therefore, by left cancellation, $y = y^2$, an idempotent in S. Conversely, suppose S is right simple and contains an idempotent. To prove that S is left cancellative, let $wy = wz$. Let $e \in S$, $e = e^2$. Since $eS = S$, for $x \in S, \exists\, y \in S \ni e\,y = x$. But $e.ey = ex$ or $ey = ex$ or $ex = x$. Let $u \in S$ be such that $w.u = e$. But $(uw)\,(uw) = u(wu)w = u(ew) = uw$, so that uw is an idempotent and hence, like e, is a left identity. Therefore, $wy = wz \Longrightarrow uwy = uwz \Longrightarrow y = z$. The proof is complete.]

(v) S is called a <u>left group</u> iff S contains an idempotent and is left simple (i.e. $Sx = S, \forall\, x \in S$). Equivalently, S is a left group iff S is right cancellative and left simple.

2.4 <u>Proposition</u>. A right group S is the direct product of a group G and a right-zero semigroup E (i.e. $\forall x, y \in E$, $x.y = y$).

<u>Proof</u>. Let E be the set of idempotents of S. Then $E \neq \phi$. Since every idempotent in S (a right group) is a left identity, E is a right-zero semigroup. Let e_o be a fixed idempotent in E. Then $S.e_o$ is a sub-semigroup of S with right identity e_o. Since S is right simple, every element a in Se_o has a

right inverse in Se_o with respect to e_o. Hence $S.e_o$ is a subgroup of S.

Let us denote Se_o by G and form the direct product $G \times E$. We define

$$\phi : G \times E \longrightarrow S$$

by $\phi((a, e)) = a.e$. Then clearly, ϕ is a homomorphism. ϕ is also 1-1, since $ae_1 = be_2$, a and $b \in G$, and $e_1, e_2 \in E$, implies $a = ae_o = a(e_1\, e_o) = (a\, e_1)\, e_o = (b\, e_2)\, e_o = b(e_2\, e_o) = be_o = b$ and hence $e_1 = e_2$, by left cancellation. Finally, ϕ is also onto, since for $a \in S$, $aS = S$ so that $\exists\, x \in S \ni ax = a$, where x easily turns out to be an idempotent. Hence ϕ is an isomorphism. Q.E.D.

2.5 Definitions. (i) An idempotent in S is called primitive if it is minimal with respect to the partial order ($\leq$) on $E(S) \equiv$ the set of idempotents of S, where $e \leq f$ (e and f idempotent) iff $ef = fe = e$.

(ii) S is called simple iff it contains no proper ideals.

(iii) S is called completely simple iff it is simple and contains a primitive idempotent.

2.6 Examples. (i) Consider Example 2.2. There in $X \times G \times Y$, the element $(x, (\phi(y, x))^{-1}, y)$ is an idempotent for each $x \in X$ and $y \in Y$ and conversely every idempotent is of this form. It is easy to see that every idempotent in $X \times G \times Y$ is primitive. Hence $X \times G \times Y$ is completely simple, since it has no proper ideals.

(ii) Here is an example of a simple semigroup with no idempotent. Consider

$$S = \left\{ \begin{bmatrix} a & 0 \\ b & 1 \end{bmatrix} \;/\; a, b \in R, \;\; a > 0, \;\; b > 0 \right\}$$

as a subset of the semigroup of all 2×2 matrices under multiplication. Then S is easily a semigroup with no idempotent. To show that S has no proper ideals, let I be an ideal of S. Then $IS \subset I$ and therefore

$I \supset \{ \begin{bmatrix} x & 0 \\ y & 1 \end{bmatrix} : x > 0, \; y > \frac{bx}{a} \}$, if $\begin{bmatrix} a & b \\ -b & 1 \end{bmatrix} \in I$.

Since $\begin{bmatrix} c & 0 \\ d & 1 \end{bmatrix} \begin{bmatrix} x & 0 \\ y & 1 \end{bmatrix} = \begin{bmatrix} cx & 0 \\ dx+y & 1 \end{bmatrix}$

is in $I \;\forall\; c > 0, \;\; d > 0, \;\; x > 0, \;\; y > \frac{bx}{a}$, it follows that $I = S$.

2.7 Proposition. Let S be a completely simple semigroup and e be a primitive idempotent of S. Then $S.e$ and $e.S$ are minimal left and right ideals respectively, such that eSe $(= Se \cap eS)$ is a group.

Proof. We first show that eS is a minimal right ideal. It is clearly a right ideal. Let I be a right ideal contained in eS. Let $a \in I \subset eS$. Then $e.a = a$. Now $SaS = S$, since S is simple. Hence there exist $x, y \in S$ such that $x.a.y = e$. Let $x' = exe$, $y' = ye$. Then $x'ay' = exeaye = e$, $ex' = x'e = x'$, $y'e = y'$. If $z' = ay'x'$, then $z'^2 = ay'(x'ay')x' = ay'ex' = ay'x' = z'$, $ez' = z'$, $z'e = z'$. Since e is primitive, $e = z' \in I$ or $eS \subset IS \subset I$. This means that eS is a minimal right ideal. Similarly, Se is a minimal left ideal of S. Hence eS is right simple and Se is left simple. Now, if $eSe = G$, then $\forall\; s \in S$, $G\,(ese) = e(Se\; se) = eSe = G$ and $(ese)G = (es.\; eS)e = eSe = G$. Hence G is a group. Clearly, $G = eS \cap Se$.

Q.E.D.

2.8 <u>Proposition</u>. Let S be a simple semigroup. Then S is completely simple iff S contains at least one minimal left ideal and at least one minimal right ideal.

<u>Proof</u>. The 'only if' part follows from Proposition 2.7. For the 'if' part, suppose that S contains at least one minimal left ideal L and one minimal right ideal R. Then R.L. = $R \cap L$ is a group (check). Let e be the identity of R.L. Suppose f is an idempotent element of S, such that $f \leq e$. Then f.e = e.f = f and therefore $f \in eSe$ = R.L. Since the identity in R.L is the only idempotent of R.L , e = f. Hence, e is a primitive idempotent and therefore, S is completely simple.

Q.E.D.

2.9 <u>Theorem</u>. A semigroup S is completely simple iff there are sets X and Y, a group G, a function $\phi : Y \times X \longrightarrow G$ such that S is isomorphic to the semigroup $X \times G \times Y$, where the multiplication is defined by $(x_1, g_1, y_1)(x_2, g_2, y_2) = x_1, g_1\, \phi(y_1, x_2)\, g_2, y_2)$.

<u>Proof</u>. The 'if' portion can easily be proved. For the 'only if' portion, let X = E(Se), Y = E(eS) and G = eSe, where e is a primitive idempotent. Since Se is a left group and eS is a right group by Lemma 2.7, X and Y are non-empty. Also G is a group. Since in a right group every idempotent is a left identity and in a left group every idempotent is a right identity, X and Y are respectively left-zero and right-zero semigroups.

Define the mapping

$$\Psi : X \times G \times Y \longrightarrow S \text{ by } \Psi(x, g, y) = x.g.y.$$

[We, of course, define multiplication in $X \times G \times Y$ by

$$(x_1,g_1,y_1)\ (x_2,g_2,y_2) = (x_1,g_1\ y_1\ x_2\ g_2,\ y_2)]$$

Then Ψ is, clearly, a homomorphism. To prove that Ψ is 1-1, we notice that

$$\begin{aligned}
(xgy)\ (exgye)^{-1} &= xgy.e.\ ((ex)\ g(ye))^{-1} \\
&= x.g.(ye).\ (ye)^{-1}.\ g^{-1}(ex)^{-1} \\
&= x\ (ex)^{-1} = x.e = x\ : \\
e(xgy)e &= (ex)\ g(ye) = ege = g\ ; \\
\text{and } (exgye)^{-1}(xgy) &= (ye)^{-1}\ g^{-1}(ex)^{-1}.(ex).gy \\
&= e.y = y.
\end{aligned}$$

This observation shows easily that Ψ is 1-1. To prove that Ψ is onto, we observe that $\{x.g.y/x \in X,\ g \in G$ and $y \in Y\}$ is an ideal ; since (i) for $s \in S$, $sxgy$ can be written as $x_1g_1y_1$ ($x_1 \in X$, $g_1 \in G$ and $y_1 \in Y$) and the reason is : sx, being an element of Se, a left group, can be written as $x_1.g'$, where $x_1 \in E(Se)$ and $g' \in eSe$ and (ii) $xgy.s$ (for $s \in S$) can similarly be written as $x_2g_2y_2$($x_2 \in X$, $g_2 \in G$, $y_2 \in Y$). Therefore, since S is simple, Ψ is onto.

Q.E.D.

2.10 <u>Ellis' Theorem</u>. We state a well-known theorem due to Ellis [2] without proof :

Let S be a locally compact space, which is algebraically a group with separately continuous multiplication. Then S is a topological group.

2.11 <u>Proposition</u> : Let S be a compact Hausdorff semigroup. Each left (right) ideal of S contains at least one minimal left (right) ideal and each minimal left (right) ideal is closed.

Proof. Let $\underline{\underline{F}} = \{$all closed left ideals of S contained in a given left ideal $I\}$. If $a \in I$, then Sa is a closed left ideal and therefore $\underline{\underline{F}}$ is non-empty. We set up a partial order in $\underline{\underline{F}}$ by inclusion. Then by compactness, every linearly ordered subset of $\underline{\underline{F}}$ has a smallest element. By Zorn's lemma, there is a minimal element I_o in $\underline{\underline{F}}$. To show that I_o is a minimal left ideal of S, let $I_1 \subset I_o \subset I$ and I_1 be a left ideal of S. But if $x \in I_1$, then $Sx \subset I_1 \subset I_o$ and Sx is a closed left ideal of S. By the minimality of I_o in $\underline{\underline{F}}$, $Sx = I_1 = I_o$. Hence I_o is a minimal left ideal of S, contained in I. Also any minimal left ideal I_o is closed since for $x \in I_o$, $Sx \subset I_o \Longrightarrow Sx = I_o$ and Sx is closed.

Q.E.D.

2.12 Theorem. Let S be a compact Hausdorff semigroup. Then the kernel K (i.e. the smallest two-sided ideal of S) of S is non-empty. Moreover, the kernel is closed and is the union of all minimal left (right) ideals, and hence it is completely simple.

Proof. Let I be a minimal left ideal of S. Let $x \in S$. Then $\{y \in I \,/\, y\,x \in J \subsetneqq I.x\}$ where J is a proper (if there is one) left ideal contained in Ix, is a proper left ideal of I and this contradicts the minimality of I. Hence Ix is a minimal left ideal of S, $\forall\, x \in S$. Clearly $K = \cup \{Ix \,/\, x \in S\}$ is a (two-sided) ideal of S. If J is an ideal of S, then $I = JI \subset J \Longrightarrow Ix \subset Jx \subset J$, $\forall\, x \in S$. Therefore, K is the kernel of S. The kernel K is clearly closed, since if $x \in K$, $S\,x\,S$ being an ideal contained in K, $K = S\,x\,S$, which is closed. The rest is clear by Prop. 2.8.

Q.E.D.

2.13 Proposition. Let S be a locally compact Hausdorff topological semigroup which is a right group. Then S is topologically isomorphic to $G \times Y$, where G is a locally compact Hausdorff topological group and Y is a locally compact Hausdorff right-zero semigroup.

Proof. Let $G = Se_o$, $e_o \in E(S)$ and $Y = E(S)$. Then G and $E(S)$ are closed in S and hence locally compact. If $\phi : G \times Y \longrightarrow S$ be defined by $\phi((g, e)) = g.e$, then as in Proposition 2.4, ϕ is an isomorphism. ϕ is also continuous. Also $\phi^{-1}(s)=(se_o,(se_o)^{-1}s)$, since $(se_o)^{-1} s(se_o)^{-1}s = (se_o)^{-1}s[e_o(se_o)^{-1}s] = (se_o)^{-1}s$ (in a right group, every idempotent is a left identity) and $(se_o)\ (se_o)^{-1}s = e_o.s = s$. Since G is a topological group by Ellis' theorem, the mapping $se_o \longrightarrow (se_o)^{-1}$ is continuous and therefore, ϕ^{-1} is clearly continuous.

Q.E.D.

2.14 Theorem. Let S be a locally compact Hausdorff topological semigroup. Suppose S has a minimal ideal K which is completely simple. Let $e \in E(K)$ = the idempotents of K, $G = eKe$, $X = E(Ke)$, $Y = E(eK)$ and let multiplication be defined in $X \times G \times Y$ by

$$(x_1,g_1,y_1)\ (x_2,g_2,y_2) = (x_1,g_1\ y_1\ x_2\ g_2,\ y_2).$$

Define :

$$\Psi : X \times G \times Y \longrightarrow K \quad \text{by} \quad \Psi(x,g,y)) = xgy$$

Then K is closed and Ψ is a topological isomorphism.

Proof. By Theorem 2.9, Ψ is an isomorphism. Now we define the mapping

$$\Psi_o : S \to X \times G \times Y \quad \text{by} \quad \Psi_o(s) = (s(ese)^{-1},ese,(ese)^{-1}s)$$

This mapping is well-defined since

(i) $s(ese)^{-1} \varepsilon S.e\, Se \subset Se = Ke$ and $[s.(ese)^{-1}][s.(ese)^{-1}]$

$= s(ese)^{-1}(ese)\ (ese)^{-1} = s(ese)^{-1}$ and therefore

$s.(ese)^{-1} \varepsilon E(Ke) = X.$

(ii) Similarly $(ese)^{-1}s \varepsilon Y$. Now we observe that

$\Psi\Psi_o : S \to S$ and for $x \varepsilon X$, $g \varepsilon G$, $y \varepsilon Y$, we have

$$\begin{aligned}\Psi\Psi_o(xgy) &= xgy\ (exgye)^{-1}.\ xgy \\ &= xgye\ (ye)^{-1}\ g^{-1}(ex)^{-1}.\ ex.g.y \\ &= xgy,\end{aligned}$$

which means that $\Psi\Psi_o$, restricted to K (every element of which can be written in the form xgy as above), is the identity mapping. Since the inversion is continuous in $eSe = eKe$, which is, by Ellis' theorem, a locally compact topological group, the mapping Ψ_o is continuous. Hence the mapping Ψ^{-1} is also continuous and consequently, Ψ is a topological isomorphism.

Also, K is closed in S, for, if $k_\alpha \varepsilon K$ and $k_\alpha \to s \not\varepsilon K$, then $\Psi\Psi_o(k_\alpha) = k_\alpha \to \Psi\Psi_o(s) \varepsilon K$, which is a contradiction.

Q.E.D.

2.15 <u>An Example</u>. Consider the semigroup S of $n \times n$ (n a positive integer) stochastic matrices $\left\{[a_{ij}] \;/\; a_{ij} \geq 0, \sum_{j=1}^{n} a_{ij} = 1, 1 \leq i \leq n\right\}$, under ordinary multiplication of matrices. Let us introduce the following topology (metric) by

$$d([a_{ij}], [b_{ij}]) = [\sum_{i,j} (a_{ij} - b_{ij})^2]^{\frac{1}{2}}.$$

Under this topology, S is a compact Hausdorff Semigroup. Every stochastic matrix A in S, where all the rows are identical, is idempotent and XA = A, for each X in S. Hence, each such stochastic matrix A is, itself, a minimal left ideal. Since for any stochastic matrix X, AX is again a stochastic matrix of the same form as A, that is, all rows of AX are again identical, the set of all stochastic matrices which are of the same form as A, that is, all rows are identical is an ideal, which is the kernel of S.

3. INVARIANT MEASURES ON SEMIGROUPS

Let S be a locally compact Hausdorff topological semigroup. Let $\underline{\underline{B}}$ be the class of all Borel sets of S, generated by the open sets. Let μ be a non-negative regular Borel measure on S, i.e.

$$\mu(B) = \inf \{\mu(V) : V \text{ open and } V \supset B\}$$

and for open V, $\mu(V) = \text{Sup} \{\mu(K) : K \text{ compact and } K \subset V\}$ and $\mu(K) < \infty$ for all compact K.

3.1 Definition. μ is called

(a) r*-invariant if $\mu(Bx^{-1}) = \mu(B)$ for each $x \in S$ and $B \in \underline{\underline{B}}$, where $Bx^{-1} = \{y \in S : yx \in B\}$.

(b) right-invariant if $\mu(Kx) = \mu(K)$ for each $x \in S$ and compact $K \subset S$.

(c) right sub-invariant if $\mu(Bx^{-1}) \leq \mu(B)$ for $x \in S$ and $B \in \underline{\underline{B}}$.

(d) right contra-invariant if $\mu(Bx^{-1}) \geq \mu(B)$ for $x \in S$ and $B \in \underline{\underline{B}}$.

(e) right infra-invariant if $\mu(Kx) \leq \mu(K)$ for $x \in S$ and compact $K \subset S$.

Similarly, different left-invariant measures are defined.

3.2 Examples. (i) Any right Haar measure on a locally compact Hausdorff topological group has all the right invariance properties.

(ii) Let S be the non-negative reals, a locally compact semigroup under addition and usual topology. Then if μ is the Lebesgue measure, μ is right sub-invariant, right-invariant, but not r*-invariant or right contra-invariant.

(iii) Let S be the positive reals ≥ 1. Then S is a locally compact semigroup under multiplication and usual topology. Let μ be the Lebesgue-Stieltjes measure induced by the function $f(x) = \log x$. Then μ is right-invariant, but not r*-invariant.

(iv) Let S be a locally compact left group, i.e. $S = E \times G$, where E is a locally compact left-zero semigroup and G is a locally compact group. Let μ_1 be non-negative regular Borel measure on E and μ_2 be a right Haar measure on G. Then the measure $\mu_1 \times \mu_2$ has all the right-invariance properties on S.

(v) Let S be the space of all ordinals less than or equal to the first uncountable ordinal Ω. Let μ be the point mass at $\{\Omega\}$. Now let $x.y = xVy$ for $x,y \in S$. Clearly, μ is r*-invariant, but not infra-invariant.

3.3 Proposition. (a) μ is right sub-invariant iff $\mu(Kx) \geq \mu(K)$ for every compact K and $x \in S$.

(b) If $\mu(S) < \infty$, then μ is right sub-invariant iff μ is right contra-invariant iff μ is r*-invariant.

Proof. (a) Suppose μ is right sub-invariant. Then $\mu(Kx) \geq \mu(Kxx^{-1}) \geq \mu(K)$, since $Kxx^{-1} \supset K$, where K is a compact set and $x \in S$.

Conversely, suppose $\mu(Kx) \geq \mu(K)$ for compact K and $x \in S$. If μ is not right sub-invariant, then one can find open V such that $\mu(Vx^{-1}) > \mu(V) + 2\varepsilon$, $\varepsilon > 0$ and for some $x \in S$. Then one can find compact $C \subset Vx^{-1}$ such that $\mu(C) > \mu(Vx^{-1}) - \varepsilon$, but $Cx \subset V$ and therefore, $\mu(Cx) \leq \mu(V) < \mu(Vx^{-1}) - 2\varepsilon < \mu(C) - \varepsilon$, a contradiction.

(b) Let $\mu(S) < \infty$. Suppose μ is right sub-invariant. Take any Borel set B

$$\mu((S-B)\, x^{-1}) \leq \mu(S-B)$$

or $$\mu(Sx^{-1}) - \mu(Bx^{-1}) \leq \mu(S) - \mu(B)$$

or $$\mu(Bx^{-1}) \geq \mu(B),$$

showing contra-invariance of μ.

Q.E.D.

3.4 Proposition. Let μ be a right sub-invariant, or right contra-invariant or right infra-invariant measure on S, a locally compact left group. Then if $S = E \times G$, where E is a locally compact left zero semigroup and G is a locally compact group, $\mu = \mu_1 \times \mu_2$, where μ_1 is a non-negative Borel measure on E and μ_2 is a right Haar measure on G.

Proof. Let us consider the representation $S = E \times G$. Let $h \in C(E)$ = the continuous functions on E with compact support such that $h \geq 0$. For $f \in C(G)$, we define $I(f) = \int h(e)\, f(g)\, \mu d\,(e,g)$. Clearly, I is a positive linear functional on $C(G)$. Also

$$I(f_{g'}) = \int h(e)\, f(gg')\, \mu d(e,\, g),$$

where $f_{g'}(x) = f(xg')$. Since the mapping $(e,g) \to h(e)\, f(gg')$ is the right translate of the mapping $(e,g) \to h(e) f(g)$ by $(e',\, g')$ (for $e' \in E$) and μ is r*-invariant, $I(f_{g'}) = I(f)$. [r*-invariance of μ follows, since if μ is right sub-invariant, then given $x \in S$ and $e = e^2 \in S$, $\exists\, y \in S \ni y.x = e$ so that for compact K, $\mu(Ky) \geq \mu(K)$ or $\mu(Kx^{-1}) \geq \mu(K)$, since $Ky \subset Kx^{-1}$.] Since I is a right invariant functional on $C(G)$, there exists a right Haar measure μ_h on G such that $\int f(g)\, \mu_h(dg) = \int h(e)\, f(g)\, \mu d(e,g)$. Therefore, given a fixed right Haar measure μ_2 on G, we can find a real number $I_o(h)$ such that $I_o(h) \int f(g)\, \mu_2(dg) = \int h(e)\, f(g)\, \mu d(eg)$ for each $f \in C(G)$. This I_o is a positive linear functional on $C(E)$ and hence there is a non-negative regular Borel measure μ_1 on E such that

$$\int h(e)\, f(g)\, \mu d(e,g) = \iint h(e)\, f(g)\, \mu_1(de)\, \mu_2(dg)$$

Hence, by an application of Stone-Weierstrass' theorem, it follows that $\mu = \mu_1 \times \mu_2$.

Q.E.D.

[Note that in a left group, sub-invariance, contra-invariance and infra-invariance are equivalent].

3.5 Definition. The support of a regular Borel measure μ on S is defined by $S_\mu = \{x \in S \,/\, \mu(V(x)) > 0$ for every open set $V(x)$ containing $x\}$. Clearly, for a non-zero measure μ, S_μ is non-empty, since otherwise $\mu(K) = 0$ for every compact set $K \subset S$. Also S_μ is closed and $\mu(S - S_\mu) = 0$, by the regularity of μ.

3.6 <u>Proposition</u>. Let μ be a r*-invariant measure on a compact semigroup S. Then S_μ is a compact left group $E \times G$, where E is a left zero semigroup and G is a compact group and μ (restricted to its support) decomposes as in 3.4.

<u>Proof</u>. We observe that $S_\mu .x \subset S_\mu \ \forall x \in S_\mu$; also $S_\mu .x = S_\mu$ since otherwise, $\exists\, y \in S_\mu - S_\mu .x$ and then there is an open set $V(y)$ containing y such that $V(y)\, x^{-1} \cap S\mu$ is empty or $\mu(V(y)\, x^{-1}) = \mu(V(y)) = 0$, a contradiction. Hence S_μ is a compact semigroup, which is left simple. Since every compact semigroup has an idempotent, S_μ is a left group. The rest follows as in 3.4.

Q.E.D.

3.7 <u>Proposition</u>. Suppose μ is a right sub-invariant measure on S such that S_μ is a left group. Then S has a kernel K, which is a closed left group and contains S_μ.

<u>Proof</u>. We observe that S_μ is a right ideal of S. If I_λ is a left ideal of S, then $S_\mu . I_\lambda$ is a left ideal of S_μ, which is a left group, $S_\mu . I_\lambda = S_\mu \subset I_\lambda$.

$$\therefore S_\mu \subset L = \bigcap \{ I_\lambda : I_\lambda \text{ is a left ideal of } S \}.$$

Clearly L is the unique minimal left ideal of S. Since $\forall\, x \in S$, Lx is a minimal left ideal of L, $L = L\,x \ \forall\, x \in S$. Hence L is also a right ideal and thus the kernel of S.

Since L is minimal, it is left simple. Also it contains at least one idempotent, namely the idempotent of S_μ. Hence L is a left group and for $e = e^2 \in L$, $L = Le = Se$, which is closed.

Q.E.D.

3.8 Theorem. Let μ be a r*-invariant measure (not necessarily finite) on S. Then S_μ is a left group iff μ is right invariant on its support.

Proof. If S_μ is a left group, then to see right invariance observe that for all compact $K \subset S_\mu$, $Kx = Kxx^{-1}$. Conversely suppose μ is r*-invariant and right invariant on its support.

Step I: Consider the mapping $\pi : S \to L$ = the space of bounded linear operators on $L_2(S, \mu)$ of norm 1, defined by

$$[\pi(s) f](x) = f(xs), \quad f \in L_2$$

To show that $\pi(x)$ is unitary, we need to show only that $\pi(x)$ is ''onto''. Let $f \in L_2$. Let $f_n(s) =$

$\sum_{i=1}^{m_n} C_{in} I_{K_{in}}(s)$, where the sets K_{in}'s are compact sets and I_K = the characteristic function of K such that

$\| f_n - f \|_2 < \frac{1}{n}$. Let $g_n(s) = \sum_{i=1}^{m_n} C_{in} I_{K_{in}x}(s)$,

so that by the right invariance of μ, $\|f_n(s) - g_n(sx)\|_2 = 0$. Since $\{g_n(sx)\}$ is Cauchy in L_2, the sequence $\{g_n(s)\}$ is Cauchy in L_2. Consequently, $\exists\, g(s) \in L_2 \ni \|g_n(s) - g(s)\|_2 \to 0$ as $n \to \infty$. It follows that $f(s) = g(sx)$ in L_2 or $\pi(x) g = f$. This proves that $\pi(x)$ is unitary $\forall\, x \in S_\mu$.

Step II. We show that $\forall\, a \in S_\mu$, $a.\,S_\mu$ is bicancellative. Since μ is r*-invariant, it follows that

$\overline{S_\mu \cdot y} = S_\mu \;\forall\, y \in S_\mu$. Also if $\pi(s_1) = \pi(s_2)$, then $as_1 = as_2$ for every $a \in S_\mu$, where $s_1 \in S_\mu$, $s_2 \in S_\mu$. To prove this, suppose $\exists\, b \in S_\mu \ni bs_1 \neq bs_2$. Then there is a continuous function f with compact support such that $f(bs_1) \neq f(bs_2)$, so that $\{x \in S_\mu \,/\, |f(xs_1) - f(xs_2)| > 0\}$ is a non-empty open set with positive μ-measure which contradicts that $f(xs_1) = f(xs_2)$ in L_2. Hence, the mapping π is an isomorphism from aS_μ into the group of unitary operators on L_2, for every $a \in S_\mu$. Therefore, $a\,S_\mu$ is bicancellative.

Step III. Let $a, b \in S_\mu$. Let K' be compact such that $\mu(K') > 0$. Then if $K = a\,K'b$,

$$\mu(a^{-1}K) \geq \mu(K'b) = \mu(K'bb^{-1}) \geq \mu(K') > 0.$$

Let K_o be compact such that $\mu(a^{-1}K \cap K_o) > 0$. Let $K_1 = K_o\,K \cup K_o$. We define : $\mu_o(B) = \mu(a^{-1}B \cap K_1)$. Then one can check that μ_o is a non-zero finite regular measure on S_μ. We now consider $\mu_o \times \mu_o$, the product measure as constructed in [24, p.152, 153]. Let $\theta(x,y) = (x, yx)$ and $\beta(x, y) = (y, x)$. Then θ is continuous and β is measure-preserving. By Fubini's Theorem [24, p. 153], we have :

$$\begin{aligned}\mu_o \times \mu_o\,(\theta(K \times K)) &= \int_K \mu_o(Kx)\,\mu_o(dx)\\ &= \int_K \mu(a^{-1}K\,x \cap K_1)\,\mu_o(dx)\\ &\geq \int_K \mu(a^{-1}K\,x \cap K_o x)\,\mu_o(dx)\\ &\geq \mu(a^{-1}K \cap K_o)\,\mu_o(K) > 0.\end{aligned}$$

Now we consider the set

$$B = a\, S_\mu\, b \times a\, S_\mu\, b - \theta(a\, S_\mu b \times a\, S_\mu b).$$

If C is compact and $C \subset B$, then for $x \in aS_\mu b$, $C_x = \{z : (x, z) \in C\}$ is contained in $aS_\mu b - aS_\mu bx$ so that

$$\mu_o(C_x) \leq \mu(a^{-1}C_x) = \mu(a^{-1}C_x\, x^{-1}\, b^{-1}) = 0,$$

since $a^{-1}\, C_x\, x^{-1}\, b^{-1} \cap S_\mu = \phi$. This means that $\mu_o \times \mu_o(C) = 0$.

Now $\mu_o \times \mu_o\ (\beta\ \theta\ (K \times K))$ is positive. Hence $\beta\ \theta\ (K \times K) \cap \theta(aS_\mu b \times aS_\mu b) \neq \phi$. Therefore, there are u, v, w and z in $aS_\mu b$ such that $(u, vu) = (zw, w)$. This means that $u = zw$, $vu = w$ or $v\, z\, w = w$ or $(vz)^2 w = (vz)w$. Since aS_μ is right-cancellative, $(vz)^2 = vz \in a\, S_\mu b \subset aS_\mu$. Let $vz = e$. Since $S_\mu .e$ is closed and $S_\mu\, e \subset S_\mu b$, $S_\mu b = S_\mu$. [Recall : $\overline{S_\mu .x} = S_\mu\ \forall\ x \in S_\mu$ and hence $S_\mu .e = S_\mu$]. Since b was arbitrarily chosen and S_μ contains the idempotent e, S_μ is left simple and therefore, a left group.

Q.E.D.

3.9 Theorem. Let μ be a r^*-invariant and ℓ^*-invariant measure on S. Then S_μ is a locally compact topological group and μ is a unimodular Haar measure on S_μ.

Proof. Observe that $\overline{S_\mu .x} = \overline{x.S_\mu} = S_\mu$ for $x \in S_\mu$. Let $a, b \in S_\mu$. We claim that $aS_\mu b$ is bicancellative. First, let $(ax_1)\ (ax_2) = (ax_1)\ (ax_3)$, where x_1, x_2 and $x_3 \in S_\mu$. Then $sx_2 = sx_3$, $s = ax_1a$. Since $\overline{S_\mu .s} = S_\mu$, there exist $\{t_\delta\}$, a net in S_μ such that $t_\delta .s \to a$. Since $t_\delta s\ x_2 =$

$t_\delta\ sx_3, ax_2 = ax_3$. Hence $aS_\mu(\supset aS_\mu b)$ is left-cancellative. Similarly, using the fact that $\overline{x\, S_\mu} = S_\mu\ \forall\ x \in S_\mu$, it follows that $S_\mu.b\ (\supset aS_\mu b)$ is right-cancellative. Now following Step III of Theorem 3.8, we can show that $aS_\mu b$ contains an idempotent e. Since $S_\mu.e = \overline{S_\mu.e} = S_\mu = \overline{e\, S_\mu} = eS_\mu$ and $aS_\mu \supset eS_\mu$, $S_\mu b \supset S_\mu.e$, we have $a\, S_\mu = S_\mu = S_\mu.b\ \forall\ a, b \in S_\mu$. Hence S_μ is a group. By Ellis' theorem, S_μ is a locally compact topological group.

Q.E.D.

3.10 Definition. Let μ, υ be two regular probability measures on S, a locally compact topological semigroup. Then the iterated integral

$$I(f) = \int \int f(ss')\ \mu(ds)\ \upsilon(ds'),$$

for $f \in C(S)$ = the continuous functions with compact support, is well-defined and defines a positive linear bounded functional on C(S). By the Riesz-representation theorem, there is a regular probability measure η on S such that $I(f) = \int f d\eta\ \forall\ f \in C(S)$. The measure η is called the convolution $\mu * \upsilon$ of μ and υ.

3.11 Lemma. For any Borel set B, the mappings $x \to \mu(Bx^{-1})$ and $x \to \mu(x^{-1}B)$ are Borél measurable.

Proof. If O is an open set, it is easy to check that

$$\mu(Ox^{-1}) = \sup_{\substack{f \in C(S) \\ 0 \le f \le 1 \\ f = 0 \text{ on } O^c}} \int f(sx)\ \mu(ds)$$

Since $\int f(sx) \, \mu(ds)$ is a continuous function of x, $\mu(Ox^{-1})$ is a lower-semicontinuous function of x and hence Borel-measurable. Let

$$\underline{\underline{F}} = \left\{ B \; / \; \mu(Bx^{-1}) \text{ is Borel measurable} \right\} .$$

Clearly, $\underline{\underline{F}}$ is a monotone class, containing all the open sets. Also, the class $\underline{\underline{F}}_0$ of all sets of the form $G \cap C$, G open and C closed and their finite disjoint unions form an algebra, since (i) $(G \cap C)^c = G^c \cup (C^c - G^c)$ (ii) $\underline{\underline{F}}_0$ is closed under finite intersection. Hence, the smallest σ-algebra containing $\underline{\underline{F}}_0$ is contained in $\underline{\underline{F}}$ so that $\underline{\underline{F}}$ contains all Borel sets.

$$Q.E.D.$$

3.12 <u>Proposition</u>. For any Borel set B,

$$\mu * \upsilon \, (B) = \int \mu(Bx^{-1}) \, \upsilon(dx) = \int \upsilon(x^{-1}B) \, \mu(dx)$$

<u>Proof</u>. First, we note that if $\underline{\underline{F}}$ is a class of non-negative continuous functions on S such that for $f_1, f_2 \in \underline{\underline{F}}$, the function $\min \left\{ f_1, f_2 \right\} \in \underline{\underline{F}}$ and P, a regular probability measure on S, then

$$\inf_{f \in \underline{\underline{F}}} \int f \, dP = \int (\inf_{f \in \underline{\underline{F}}} f) \, dP.$$

Now let us define the measure λ by

$$\lambda(B) = \int \mu(Bx^{-1}) \, \upsilon(dx).$$

Using the regularity of the measure μ_x (defined by $\mu_x(B) = \mu(Bx^{-1})$) and the upper-semi-continuity of the mapping $x \to \mu(Kx^{-1})$ for compact K, it is easy to show that given $\epsilon > 0$, $\exists$ compact K_ϵ such that $\lambda(K_\epsilon) > 1 - \epsilon$. This means that $\lambda(C) = \sup \left\{ \lambda(K) : K \subset C, \right.$

K compact$\}$, whenever C is closed.

Finally, if K is a compact set, then

$$\lambda(K) = \int \mu(Kx^{-1})\, \upsilon(dx)$$

$$= \int \inf_{\substack{f \in C(S) \\ f=1 \text{ on } K}} \left[\int f(sx)\, \mu(ds)\right] \upsilon(dx)$$

$$= \inf_{\substack{f \in C(S) \\ f = 1 \text{ on } K}} \int \int f(sx)\, \mu(ds)\, \upsilon(dx)$$

$$= \inf_{\substack{f \in C(S) \\ f=1 \text{ on } K}} \int f(s)\, \mu * \upsilon(ds) = \mu * \upsilon(K),$$

since $\mu * \upsilon$ is regular. Hence, for any closed set C, $\lambda(C) = \mu * \upsilon(C)$. Therefore, for open O, $\lambda(O) = 1 - \lambda(O^c) = 1 - \mu * \upsilon(O^c) = \mu * \upsilon\ (O)$.

Now

$$\mu * \upsilon(B) = \inf_{\substack{B \subset O \\ O \text{ open}}} \mu * \upsilon(O) = \inf_{\substack{B \subset O \\ O \text{ open}}} \lambda(O) \geq \lambda(B).$$

Also $$\mu * \upsilon(B) = \sup_{\substack{K \subset B \\ K \text{ compact}}} \mu * \upsilon(K) = \sup_{\substack{K \subset B \\ K \text{ compact}}} \lambda(K) \leq \lambda(B).$$

Hence $\mu * \upsilon\ (B) = \lambda(B)$ for all Borel sets B.

Q.E.D.

3.13 Proposition. For $\mu, \upsilon \in P(S)$ = the set of regular probability measures on S,

$$\overline{S_\mu \cdot S_\upsilon} = S_{\mu * \upsilon}$$

Proof. Let $s \in S_\mu$, $t \in S_\upsilon$. If $s.t. \notin S_{\mu * \upsilon}$, then $\exists$ open $V(st)$ containing $st \ni \mu * \upsilon\ (V(st)) = 0$. But $\mu(V(st)\ t^{-1}) \geq \mu(V(s)) > 0$. By lower-semicontinuity of the mapping $x \to \mu(V(st)\ x^{-1})$, there is an open set $N(t)$

containing t such that $\mu(V(st)y^{-1}) > 0 \ \forall\ y \in N(t)$. Hence, since $\upsilon(N(t)) > 0$, $\int \mu(V(st)\ y^{-1})\ \ \upsilon(dy) > 0$, a contradiction. Hence $\overline{S_\mu \cdot S_\upsilon} \subset S_{\mu * \upsilon}$. Conversely, let $z \in S_{\mu * \upsilon}$. If $z \notin \overline{S_\mu . S_\upsilon}$, then there is an open set $N(z)$ containing z such that $N(z) \cap S_\mu . S_\upsilon = \phi$ or $N(z)\ y^{-1} \cap S_\mu = \phi \ \forall\ y \in S_\upsilon$ or $\int \mu(N(z)y^{-1})\ \ \upsilon(dy) = 0$. Then $\mu * \upsilon\ (N(z)) = 0$, a contradiction.

Q.E.D.

3.14 <u>Proposition</u>. Suppose $\mu, \upsilon \in P(S)$ and

$$\mu = \mu * \upsilon = \upsilon * \mu.$$

Then for every Borel set B,

$$\mu(Bx^{-1}\ y^{-1}) = \mu(Bx^{-1})$$

and

$$\mu(y^{-1}\ x^{-1}\ B) = \mu(x^{-1}B)$$

for $x \in S_\mu$, $y \in S_\upsilon$.

<u>Proof</u>. We will only prove the first equality since the second one will follow similarly. First, we observe that for every bounded measurable function $f(s)$ on S, we have

$$\int f(s)\ \mu(ds) = \int \int f(st)\ \mu(ds)\ \upsilon(dt)$$
$$= \int \int f(st)\ \upsilon(ds)\ \mu(dt),$$

since $\mu = \mu * \upsilon = \upsilon * \mu$. Let K be any compact set and $x \in S_\mu$. Let $\mu(Kx^{-1}) = a$. Let ε be an arbitrary positive number. Then by the regularity of μ_x (where $\mu_x(B) = \mu(Bx^{-1})$, we can find open sets W and O and a closed set V such that $O \supset V \supset W \supset K$ and $\mu(Ox^{-1}) < a + \varepsilon$. Let $A = \{s: \mu(Vs^{-1}) \geq a+\varepsilon\}$. Then A^c is open. Since $x \in A^c \cap S_\mu$, $\mu(A^c) > 0$. Let

$g(s) = \max \{\mu(Vs^{-1}) - a - \varepsilon, 0\}$. Then

$$\int [\int g(st) \ \upsilon(ds) - g(t)] \ \mu(dt) = 0.$$

Since $g(t) \leq \int g(st) \ \upsilon(ds)$, it follows that for some Borel set E with $\mu(E) = 0$ and for all t in $S-E$, we have $g(t) = \int g(st) \ \upsilon(ds)$. Let $y \in A^c - E$. Then $g(y) = 0$ and therefore, $g(sy) = 0$ for υ-almost all s. Since W is open and $W \subset V$, $\mu(Wx^{-1} s^{-1}) \leq a + \varepsilon \ \forall s \in S_\upsilon$. Since $K \subset W$, $\mu(Kx^{-1} s^{-1}) \leq \mu(Kx^{-1}) + \varepsilon \ \forall s \in S_\upsilon$. Since ε is arbitrary and $\mu(Kx^{-1}) = \int \mu(Kx^{-1} s^{-1}) \ \upsilon(ds)$, we have $\mu(Kx^{-1}) = \mu(Kx^{-1} s^{-1})$ for υ-almost all s and hence for all $s \in S_\upsilon$. The proposition now follows from the regularity of μ.

Q.E.D.

3.15 <u>Theorem</u>. Suppose $\mu \in P(S)$ and $\mu = \mu * \mu$. Then S_μ is a closed completély simple semigroup.

<u>Proof</u>. Since $\overline{S_\mu . S_\mu} = S_\mu$ (by proposition 3.13), S_μ is a closed sub-semigroup of S. By Proposition 3.14, we have

$$\left.\begin{aligned} \mu(Bx^{-1} y^{-1}) &= \mu(Bx^{-1}) \\ \text{and} \quad \mu(y^{-1}x^{-1}B) &= \mu(x^{-1}B) \end{aligned}\right\} \quad \begin{aligned} &\text{for } x,y \in S_\mu \text{ and} \\ &\text{Borel set } B \subset S_\mu \end{aligned}$$

Now consider for any $a \in S_\mu$, the measure μ_a (defined by $\mu_a(B) = \mu(Ba^{-1})$) on its support $\overline{S_\mu . a}$. For any $z \in S_\mu$,

$$\mu_a(B(za)^{-1}) = \mu(Ba^{-1}z^{-1}a^{-1}) = \mu(Ba^{-1}(az)^{-1}) = \mu_a(B) ;$$

also if $z \in \overline{S_\mu . a}$ and $z_\delta a$ is a net converging to z, $z_\delta \in S_\mu$, then for compact K,

$$\mu_a(Kz^{-1}) \geq \limsup \ \mu_a(K(z_\delta a)^{-1}) = \mu_a(K).$$

On the other hand, given $\varepsilon > 0$, if U is open, $U \supset K$ and $\mu_a(U - K) < \varepsilon$, then

$$\mu_a(Kz^{-1}) \leq \mu_a(Uz^{-1}) \leq \liminf \mu_a(U(z_\delta a)^{-1})$$
$$= \mu_a(U) \leq \mu_a(K) + \varepsilon .$$

Hence μ_a is r*-invariant for every compact set $K \subset \overline{S_\mu . a}$. By regularity of the measure μ, it follows easily that μ_a is r*-invariant for every Borel set $B \subset \overline{S_\mu . a}$. Now since $\mu(y^{-1}x^{-1}B) = \mu(x^{-1}B)\ \forall\ x,y \in S_\mu$ and Borel $B \subset S_\mu$, by dual arguments as above, the measure ${}_a\mu$ (defined by ${}_a\mu(B) = \mu(a^{-1}B)$) is ℓ*-invariant on $\overline{aS_\mu}$. We consider next the measure m (defined by $m(B) = \mu(a^{-1}Ba^{-1})$) on its support $Q = \overline{a S_\mu a}$. This measure is regular and both r*- and ℓ*-invariant. Therefore, it follows from Theorem 3.9 that the support $\overline{a S_\mu a}$ of m is a compact topological group. Let $\overline{S_\mu . a} = G$ and $\overline{aS_\mu} = H$. Since G is the support of μ_a, an r*-invariant measure, $\overline{Gy} = G\ \forall\ y \in G$. Now Gy as a left ideal of G and Q as a right ideal of G, intersect each other and hence since $Q \cap Gy$ is a left ideal of Q, a group, $Q \cap Gy = Q$ or $Q \subset Gy$. Therefore, the identity e of Q is in Gy. Then $Ge \subset Gy$. But $Ge = \overline{Ge} = G$. Hence $Gy = G\ \forall\ y \in G$. It follows that G is a left group. Therefore, $\forall\ a \in S_\mu$, $S_\mu . a = \overline{S_\mu . a}$ is a minimal left ideal of S_μ. Similarly, aS_μ is a minimal right ideal of S_μ. Hence by Proposition 2.8, S_μ has a kernel (the union of all minimal left ideals), which is completely simple. The kernel $K \supset S_\mu . S_\mu$ and the kernel K is closed by Theorem 2.14. Hence $K = S_\mu (= \overline{S_\mu . S_\mu})$.

Q.E.D.

3.16 <u>Theorem</u>. Let $\mu = \mu * \mu$. Let e be an idempotent of S_μ (which is completely simple). Then $\exists\ \mu_1 \in P(E(S_\mu .e))$, μ_2, the Haar measure of $eS_\mu e$, a compact group and $\mu_3 \in P(E(eS_\mu))$ such that $\mu = \mu_1 * \mu_2 * \mu_3$. Conversely, if μ_2 is the Haar probability measure on a compact subgroup of S and $\mu_1, \mu_3 \in P(S)$ such that $S_{\mu_3} . S_{\mu_1} \subset S_{\mu_2}$, then $\mu_1 * \mu_2 * \mu_3$ is idempotent.

<u>Proof</u>. We know that S_μ is a completely simple semigroup. Also e is an idempotent of S_μ and δ_e, the point mass at $\{e\}$, then $\mu * \delta_e * \delta_e * \mu = \mu$, since

$$\begin{aligned}\mu * \delta_e * \mu\ (B) &= \int \mu * \delta_e(Bx^{-1})\ \mu(dx)\\ &= \int \mu(Bx^{-1}e^{-1})\ \mu(dx)\\ &= \int \mu(Bx^{-1})\ \mu(dx), \text{ by Prop. 3.14}\\ &= \mu(B).\end{aligned}$$

Also $\mu * \delta_e * \mu = (\mu * \delta_e) * (\delta_e * \mu)$, since $\delta_e * \delta_e = \delta_e$. Now $\mu * \delta_e (= \mu_e)$ is an r^*-invariant probability measure with support $S_\mu .e$ and $\delta_e * \mu$ is an ℓ^*-invariant probability measure with support eS_μ. We know that $S_\mu .e$ is a left group (by Theorem 3.15, since any r^*-invariant probability measure m is idempotent and $\overline{S_m .x} = S_m\ \forall\ x \in S_m$ so that S_m is completely simple and then a left group). Similarly, $e.S_\mu$ is a right group. Also $S_\mu .e$ is topologically isomorphic to $E(S_\mu e) \times eS_\mu e$ by the mapping $\phi : (x,y) \to x.y$ and there exists $\mu_1 \in P(E(S_\mu e))$ and μ_2, the normed Haar measure of $aS_\mu e$, a compact group such that

$\forall f \in C(E(S_\mu e) \times eS_\mu e)$,

$\int \int f(x,y)\ \mu_1(dx)\ \mu_2(dy) = \int f.\ \phi^{-1}\ d\mu_e$,

by Proposition 3.4.

But for Borel set $A \subset E(S_\mu e)$, $B \subset eS_\mu e$,

$$\mu_1 * \mu_2\ (\phi(A \times B)) = \int \mu_1(\phi(A \times B)\ x^{-1})\ \mu_2(dx)$$

$$= \mu_1(A)\ \mu_2(B) = \mu_1 \times \mu_2(A \times B),$$

a left group being right cancellative and ϕ being 1-1. Hence, by using Stone-Weierstrass' Theorem, one can show that

$\forall f \in C(E(S_\mu e) \times eS_\mu e)$,

$\int \int f(x,y)\ \mu_1(dx)\ \mu_2(dy) = \int f \circ \phi^{-1}\ d\mu_1 * \mu_2$.

This means that $\mu_e = \mu * \delta_e = \mu_1 * \mu_2$. Similarly, ${}_e\mu = \delta_e * \mu = \mu_2 * \mu_3$, for some $\mu_3 \in P(E(eS_\mu))$. Therefore,

$$\mu = \mu_e * {}_e\mu = (\mu_1 * \mu_2) * (\mu_2 * \mu_3)$$

$$= \mu_1 * \mu_2 * \mu_3 .$$

Conversely, if μ_2 is the Haar measure on a compact subgroup of S and μ_1, $\mu_3 \in P(S)$ with $S_{\mu_3} \cdot S_{\mu_1} \subset S_{\mu_2}$, then $\mu_3 * \mu_1 \in P(S_{\mu_2})$. Hence $\mu_2 * \mu_3 * \mu_1 = \mu_2$. Then

$$(\mu_1 * \mu_2 * \mu_3) * (\mu_1 * \mu_2 * \mu_3)$$

$$= \mu_1 * (\mu_2 * \mu_3 * \mu_1 * \mu_2) * \mu_3$$

$$= \mu_1 * \mu_2 * \mu_3.$$

Q.E.D.

3.17 <u>Theorem</u>. Let $\mu \in P(S)$, $\mu = \mu * \mu$. Then S_μ is completely simple. Let $X = E(S_\mu e)$, $G = eS_\mu e$, $Y = E(eS_\mu)$, where $e = e^2 \in S_\mu$. The measure μ decomposes on $X \times G \times Y$ as a product measure $\mu = \mu_1 \times \mu_2 \times \mu_3$, where $\mu_1 \in P(X)$, $\mu_2 =$ the normed Haar measure of G and $\mu_3 \in P(Y)$.

Proof. From Theorem 3.16, we know that $\mu = \mu_1 * \mu_2 * \mu_3$, where μ_1, μ_2 and μ_3 are as stated in the Theorem. Since S_μ is completely simple, we know that the mapping $\phi : X \times G \times Y \to S_\mu$ defined by $\phi(x,g,y) = x.g.y$ is a topological isomorphism. Also the mapping $\phi_o : X \times G \to S_\mu e$ (= the support of $\mu_1 * \mu_2$) defined by $\phi_o(x,g) = x.g$ is a topological isomorphism.

Since $\mu = \mu_1 * \mu_2 * \mu_3$, we have $\forall f \in C(S_\mu)$,

$$\int fd\,\mu = \int\int\int f(xgy)\,\mu_1(dx)\,\mu_2(dg)\,\mu_3(dy).$$

We notice that for Borel set $A \subset X$, $B \subset G$, $C \subset Y$,

$$\mu_1 * \mu_2 * \mu_3\,(\phi(A \times B \times C))$$

$$= \int \mu_1 * \mu_2\,(\phi(A \times B \times C)\,y^{-1})\,\mu_3(dy)$$

$$= \int_C \mu_1 * \mu_2(\phi_o(A \times B))\,\mu_3(dy),$$

since $w \in \phi(A \times B \times C)y^{-1} \Longleftrightarrow w.y \in \phi(A \times B \times C)$ and if $w = x.g$, $x \in X$, $g \in G$, then since ϕ is 1-1, $y \in C$ and $w = x.g \in \phi_o(A \times B)$, $x \in A$, $g \in B$. Since

$\mu_1 * \mu_2\,(\phi_o(A \times B)) = \mu_1(A)\,\mu_2(B)$, we have

$\mu_1 * \mu_2 * \mu_3(\phi(A \times B \times C)) = \mu_1(A)\,\mu_2(B)\,\mu_3(C)$.

Now, by using Stone-Weierstrass' Theorem, it follows that

$\int fd\,\mu = \int\int\int f \circ \phi^{-1}\,d\,\mu_1\,d\mu_2\,d\,\mu_3$.

Q.E.D.

3.18 Theorem. Let $\mu \in P(S)$, μ be r*-invariant. Then S_μ is a left group. Let S_μ be represented as $E(S_\mu) \times e\,S_\mu$, $(e = e^2 \in S_\mu)$. Then μ on S_μ decomposes as $\mu_1 \times \mu_2$, where $\mu_1 \in P(E(S_\mu))$ and μ_2 is the normed Haar measure of the compact subgroup eS_μ.

Proof. The theorem easily follows from 3.15 and 3.4, since every r*-invariant probability measure is an idempotent probability measure.

Q.E.D.

Comments on the results of section 3.

Some of the discussions on invariant measures are given by Berglund and Hofmann in [2]. The r*-invariant measures were studied in certain locally compact semigroups (satisfying a compactness condition) by Argabright [1]. See also Rosen [64]. These measures were also considered by Berglund and Hofmann [2]. The authors proved Theorem 3.18 in [53] and Theorem 3.8 in [55]. At present, it is not known if Theorem 3.18 is still valid for infinite measures.

The Choquet-Deny Convolution equation in Prop. 3.14 was considered first in [7] for abelian groups, then in [75, 76] by Tortrat for non-abelian groups and certain simple semigroups and then in [46] by Mukherjea on general locally compact semigroups.

Idempotent probability measures were first studied in [31] and then again, independently, in [79], on compact groups. In the case of locally compact groups, they were characterized as normed Haar measures on compact subgroups for the first time in [32] and then, independently, by Pym in [61], by Heyer in [27]. Heble and Rosenblatt in [24] and Pym in [61] (independently) characterized these measures on compact semigroups. The characterization of these measures on general locally compact semigroups, as given in Theorem 3.15, 3.16 and 3.17, are results of Mukherjea and Tserpes in [53]. These results along with Theorem 3.9 were then proven by the authors in the more general context of semitopological semigroups (where the multiplication is separately continuous) in [54]. In [47], Mukherjea studied infinite measures (which are like idempotent probability measures) satisfying $P(B) = \int P(Bx^{-1})\, P(dx)$ on locally compact semigroups and showed, among other things, that such measures cannot exist on locally compact groups.

The books of Grenander [22], Parthasarathy [59] and Rosenblatt [68] contain useful discussions of idempotent probability measures in the context of topological groups and compact semigroups.

4. Limit Theorems for Probability Measures

Let S be a locally compact Hausdorff semigroup and $C(S)$ be the set of continuous functions (real-valued) with compact support. Then by Banach-Alaoglu theorem, it follows that the set $\{\mu \mid \mu$ is a regular Borel measure and $\mu(S) \leq 1\}$ is a compact set in the weak-star topology ++
(i.e. the topology where $\mu_\delta \to \mu$ iff $\int f\, d\mu_\delta \to \int f d\mu \;\forall f \in C(S)$). In this topology, the set $P(S) = \{\mu / \mu$ is a regular probability measure$\}$ need not be compact for instance, if $S = [0, \infty)$ with usual topology and multiplication and μ_n be the point mass at $\{n\} \in S$, then μ_n converges to the zero measure. However, this set is compact when S is compact.

4.1 <u>Proposition</u>. Let S be a compact Hausdorff semigroup. Then $P(S)$ is compact in the weak-star topology. [We present a complete proof, as in that of the Banach-Alaoglu Theorem.]

++ The weak-star topology is replaced by weak topology when we consider all bounded continuous functions.

<u>Proof</u>. Let $I = [0, 1]$ with usual topology. Consider the product $X = \prod_{\substack{f \in C(S) \\ 0 \leq f \leq 1}} I_f$, $I_f \equiv I$. Then by Tychonoff Theorem, X is compact in the usual product topology. Let ϕ be the mapping from $P(S)$ into X defined by $\phi(\mu)$, a point in X whose component is the f-th co-ordinate space I_f is $x_f = \int f \, d\mu$. Clearly ϕ is a 1-1, continuous and open mapping. We claim that $\phi(P(S))$ is a closed subset of X. To prove our claim, let $x \in X$ be a limit point of $\phi(P(S))$. Then $\forall f \geq 0$, $x_f \geq 0$. Given any two functions f_1, f_2 with $0 \leq f_1, f_2$, $f_1 + f_2 \leq 1$ and any $\varepsilon > 0$, there is a measure $\mu \in P(S) \ni \phi(\mu)$ is in the neighborhood

$$\{x' : |x'_{f_1} - x_{f_1}| < \varepsilon , |x'_{f_2} - x_{f_2}| < \varepsilon \text{ and } |x'_{f_1+f_2} - x_{f_1+f_2}| < \varepsilon\}$$

of x. Since $\varepsilon > 0$ is arbitrary,

$$x_{f_1} + x_{f_2} = x_{f_1+f_2}.$$

Similarly, $x_{\delta f} = \delta . x_f$, for $\delta \geq 0$, $0 \leq \delta f$, $f \leq 1$.

Let $I(f) = x_f$, for $0 \leq f \leq 1$, $f \in C(S)$.

For any $f \in C(S)$ and $f \geq 0$, define

$I(f) = \frac{1}{\delta} I(\delta f)$, where $\delta \geq 0$ is $\ni 0 \leq \delta f \leq 1$.

For any $f \in C(S)$, define $I(f) = I(f^+) - I(f^-)$,

where $f^+ = \max \{f, 0\}$ and $f^- = -f + f^+$.

Then I is a positive linear functional on $C(S)$ and for $f \in C(S)$,

$$|I(f)| = |I(f^+) - I(f^-)|$$
$$\leq \max\{I(f^+), I(f^-)\} \leq ||f||$$

(the usual supremum norm) and therefore $||I|| \leq 1$. By the Riesz-representation theorem, there is a regular probability measure μ_o on S (since $I(1) = 1$) such that

$$I(f) = \int f d\mu_o$$

This means that $\phi(\mu_o) = x$ or $\phi(P(S))$ is a closed subset of the compact space X. Hence $\phi(P(S))$ is compact and therefore $P(S)$ is compact.

Q.E.D.

4.2 Proposition. The convolution operation for regular probability measures on S (not necessarily compact) is jointly continuous in the weak-star topology and therefore $P(S)$ is a topological semigroup in this topology.

Proof. Let $\mu_\delta \in P(S)$ and $\mu \in P(S)$ be such that $\mu_\delta \to \mu$ in the weak-star topology. Then given $\varepsilon > 0$, $\exists$ compact set K such that $\mu(K) > 1 - \varepsilon$. Let $K \subset O$, O open and $\bar{O}$ compact. Let $f \in C(S)$ be such that $0 \leq f \leq 1$, $f(x) = 1$ if $x \in K$, $f(x) = 0$, $x \notin O$. Then since $\int f d\mu_\delta \to \int f d\mu$, it follows that eventually $\mu_\delta(K) > 1 - \varepsilon$. If g is any bounded continuous function on S, then $|\int g.f\, d\mu_\delta - \int g\, d\mu_\delta| < \varepsilon$ eventually and $|\int gf d\mu - \int g d\mu| < \varepsilon$. Also since $g.f \in C(S)$, $\int g\, f\, d\mu_\delta \to \int g.f\, d\mu$. This means that $\int g d\mu_\delta \to \int g\, d\mu\ \forall$ bounded continuous function g.

Now to prove joint continuity of the convolution operation, let $\{\mu_\delta\}$ and $\{\upsilon_\delta\}$ be two nets of probability measure in $P(S)$ such that $\mu_\delta \to \mu \in P(S)$ and $\upsilon_\delta \to \upsilon \in P(S)$. Given $\varepsilon > 0$, by above argument we can find a compact set K such that,

$$\forall\delta,\ \mu_\delta(K) > 1 - \varepsilon,\ \upsilon_\delta(K) > 1 - \varepsilon,\ \mu(K) > 1 - \varepsilon$$
$$\text{and } \upsilon(K) > 1 - \varepsilon.$$

Then given any function $h \in C(S)$, using a compactness type of argument, we can show easily that

$$\left|\int_K \int_K h(st)\ \mu_\delta(ds)\ \upsilon_\delta(dt) - \int_K \int_K h(st)\ \mu(ds)\ \upsilon(dt)\right|$$

is less than ε, eventually. It follows that

$$\left|\iint h(st)\ \mu_\delta(ds)\upsilon_\delta(dt) - \iint h(st)\ \mu(ds)\ \upsilon(dt)\right|$$

can be made arbitrarily small eventually. This means that $\mu_\delta * \upsilon_\delta \to \mu * \upsilon$.

Q.E.D.

4.3 <u>Proposition</u>. Let $\mu \in P(S)$, where S is a compact Hausdorff semigroup. Then the averaged convolution sequence

$$\left(\frac{1}{n}\right)\sum_{j=1}^{n} \mu^j .$$

converges to a probability measure $\lambda \in P(S)$. Also

$$\mu * \lambda = \lambda * \mu = \lambda = \lambda * \lambda.$$

<u>Proof</u>. Let $\mu_n = \left(\frac{1}{n}\right)\sum_{j=1}^{n} \mu^j$. Then $\{\mu_n\}$ is a sequence in the compact semigroup $P(S)$. If λ_1 is a limit-point of this sequence, then $\mu * \lambda_1 = \lambda_1 = \lambda_1 * \mu$; because $\mu_n = \mu * \mu_n + \frac{1}{n}[\mu - \mu^{n+1}]$. Hence $\mu^n * \lambda_1 = \lambda_1 = \lambda_1 * \mu^n$, so that if λ_2 is another limit-point, $\lambda_2 * \lambda_1 = \lambda_1 = \lambda_1 * \lambda_2$. Similarly, $\lambda_2 * \lambda_1 = \lambda_2 = \lambda_1 * \lambda_2$. Hence $\lambda_1 = \lambda_2$ and the theorem follows.

Q.E.D.

4.4 Proposition. Let μ be a regular probability measure on a compact Hausdorff semigroup S, generated by the support of μ. Then if K is the kernel of S and O is any open set containing K, the sequence $\{\mu^n(O)\}$ converges to 1 as $n \to \infty$.

Proof. Since $S = \overline{\bigcup_{n=1}^{\infty} (S_\mu)^n}$, there is a positive integer k such that $N(x) \cap (S_\mu)^k \neq \phi$, where $N(x)$ is a neighborhood of $x \in K$ and $N(x) \subset O$. This means that there are $x_i \in S_\mu$, $1 \leq i \leq k$ such that for suitable open sets $N(x_i)$ containing x_i,

$$N(x_1)\ N(x_2)\ \dots\ N(x_k) \subset N(x).$$

We can and do assume that $N(x)$ has been chosen so that $SN(x)\,S \subset O$, since S is compact.

Consider now the infinite sequence space S^∞ and the product measure P induced by μ. Then the co-ordinate projections $\{X_n\}_{n=1}^{\infty}$ are a sequence of independent S-valued random variables on S^∞ with identical distribution μ.

Clearly $\sum_{n=0}^{\infty} P[X_{nk+i} \in N(x_i),\ 1 \leq i \leq k] = \infty$.

By Borel-Cantelli Lemma,

$$P[\bigcap_{j=0}^{\infty} \bigcup_{n=j}^{\infty} \{X_{nk+i} \in N(x_i),\ 1 \leq i \leq k\}] = 1.$$

Hence given $\varepsilon > 0$, $\exists\, j_o \ni j \geq j_o \Rightarrow$

$$P[\bigcup_{n=0}^{j} \{X_{nk+i} \in N(x_i),\ 1 \leq i \leq k\}] > 1 - \varepsilon.$$

Since

$$\bigcup_{n=0}^{j} \{x_{nk+i} \in N(x_i),\ 1 \le i \le k\}$$

$$\subset \{x_1\, x_2 \ldots x_{(j+1)k} \in SN(x_1) \ldots N(x_k)S \subset O\},$$

it follows that for $n \ge (j_o + 1)k$,

$$\mu^n(O) > 1 - \varepsilon.$$

[We note that the set $\{x_1\, x_2 \ldots x_m \in B\}$ is a measurable set in the product space S^∞ whenever B is a compact G_δ and given any open set O containing K, we can find open O_1, and a compact G_δ set O_2 such that $K \subset O_1 \subset O_2 \subset O$. Thus the difficulty in the measurability of the set $\{x_1\, x_2 \ldots x_m \in O\}$ can be avoided.]

4.5 Proposition. If S is a compact Hausdorff semigroup generated by the support of $m \in P(S)$ and $\mu * m = \mu$, for some $\mu \in P(S)$, then $\mu = \mu^2$.

Proof. Let K be the kernel of S, which is closed. We claim that $S_\mu \subset K$. If the claim is false, there exists an $x \in S_\mu$ such that $x \notin K$. Then there is an open set $N(x)$ containing x and an open set $O \supset K$ such that $N(x) \cap O = \phi$. Now for $z \in S$, $K \subset z^{-1}O$, since $z\,K \subset K \subset O$. Hence by Proposition 4.4, $m^n(z^{-1}O) \to 1$ as $n \to \infty$, for every $z \in S$. Since $z^{-1}O \cap z^{-1}N(x) = \emptyset$, $m^n(z^{-1}N(x)) \to 0$ as $n \to \infty$ for every $z \in S$. But $\mu * m = \mu$ and therefore $\mu * m^n = \mu$ and $\mu(N(x)) = \int m^n(z^{-1}N(x))\, \mu(dz)$ which goes to 0 as $n \to \infty$. This is a contradiction since $x \in S_\mu$. Hence $S_\mu \subset K$. Now by Proposition 4.3, there is a $\lambda \in P(S)$ such that the averaged convolution sequence

$\left(\frac{1}{n}\right)\sum_{k=1}^{n} m^k \to \lambda$ in the weak-star topology and $\lambda = \lambda^2$, $m * \lambda = \lambda * m$

$= \lambda$. Now since $\mu * (\frac{1}{n} \sum_{k=1}^{n} m^k) = \mu$, $\mu * \lambda = \mu$. Since $m^n * \lambda = \lambda * m^n = \lambda$, $S_m^n \cdot S_\lambda = S_\lambda \cdot S_m^n = S_\lambda$ for each positive integer n. This means that S_λ is an ideal of

$S = \overline{\bigcup_{n=1}^{\infty} S_m^n}$. Since λ is an idempotent probability measure, S_λ is also simple and therefore $S_\lambda = K \supset S_\mu$. Also $\lambda(y^{-1}x^{-1}B) = \lambda(x^{-1}B)$ for $x \in K$, $B \subset K$, $y \in K$. Since

$\mu * \lambda = \mu$, for $x \in S_\mu$, $B \subset S_\mu$,

$$\mu(x^{-1}B) = \int \lambda(y^{-1}x^{-1}B)\,\mu(dy)$$
$$= \lambda(x^{-1}B)$$

and therefore,
$$\mu(B) = \int \lambda(x^{-1}B)\,\mu(dx)$$
$$= \int \mu(x^{-1}B)\,\mu(dx)$$
$$= \mu^2(B).$$

Q.E.D.

4.6 Corollary. Let S be a compact Hausdorff semigroup generated by the support of $m \in P(S)$. For $\mu \in P(S)$, $\mu * m = m * \mu = \mu$ iff $\left(\frac{1}{n}\right) \sum_{k=1}^{n} m^k$ converge to μ in the weak-star topology.

The proof is omitted.

4.7 Proposition. Let S be a locally compact Hausdorff semigroup generated by the support of $m \in P(S)$. Let $\mu \in P(S)$ and $\mu * m = m * \mu = \mu$. Then

(i) μ^2 is idempotent;

(ii) For $x \in S_\mu$, $S_\mu . x$ is a left group and $x . S_\mu$ is a right group;

(iii) If S_μ is right cencellative (or left-cancellative), then $\mu = \mu^2$;

(iv) If $m^n(O) \to 1$ as $n \to \infty$ for open $O \supset S_\mu$, then $\mu = \mu^2$;

(v) If $m^k(S_\mu) > 0$ for some positive integer k, then $\mu = \mu^2$.

Proof. First we prove (ii). By 3.14, for any Borel set $B \subset S_\mu$, $x \in S_\mu$ and $y \in S$, we have

$$\begin{aligned} \mu(Bx^{-1} y^{-1}) &= \mu(Bx^{-1}) \\ \mu(y^{-1} x^{-1} B) &= \mu(x^{-1}B) \end{aligned} \qquad (1)$$

Since $\mu * m = m * \mu = \mu$, we have $\mu * m^n = m^n * \mu = \mu$ for all n. Therefore,

$$\overline{S_\mu \cdot S_m^n} = \overline{S_m^n S_\mu} = S_\mu .$$

Since $S = \overline{\bigcup_{n=1}^{\infty} S_m^n}$, S_μ is an ideal of S. From the equalities in (1), it is easy to check that for each $x \in S_\mu$, the measure μ_x ($= \mu * \delta_x$, δ_x the point mass at x) is r*-invariant on its support $\overline{S_\mu . x}$. Hence by 3.6, $\overline{S_\mu . x}$ is a left group. If $e = e^2 \in \overline{S_\mu . x}$, then $S_\mu . e \subset \overline{S_\mu . x}$. Also $S_\mu . e \supset \overline{S_\mu . x}.e = \overline{S_\mu . x}$. Therefore $\overline{S_\mu . x} = S_\mu . e$. If $z \in S_\mu$, then $z.x \in S_\mu . e$ and therefore, $S_\mu . e = S_\mu . e.z.x \subset S_\mu . x$. Hence $S_\mu . x = \overline{S_\mu . x} = S_\mu . e$, a left group. Similarly $x.S_\mu$ is a right group, for $x \in S_\mu$.

To prove (i), for $x \in S_\mu$, $B \subset S_\mu$,

$$\begin{aligned} \mu^2(Bx^{-1}) &= \int \mu(Bx^{-1} y^{-1})\, \mu(dy) \\ &= \int \mu(Bx^{-1})\, \mu(dy) = \mu(Bx^{-1}) \end{aligned} \qquad (2)$$

or

$$\begin{aligned} \mu^3(B) &= \int \mu^2(Bx^{-1})\, \mu(dx) \\ &= \int \mu(Bx^{-1})\, \mu(dx) \\ &= \mu^2(B) \end{aligned}$$

Hence $\mu^2 = \mu^3$ or μ^2 is idempotent.

[We note here that if the equalities in (1) hold for all $x \in S$, $y \in S$, then $\mu^2(Bx^{-1}) = \mu(Bx^{-1})$ for all $x \in S$ and so

$$\mu^2(B) = \int \mu^2(Bx^{-1})\, m(dx) = \int \mu(Bx^{-1})\, m(dx) = \mu(B) \text{ or } \mu = \mu^2.$$

So the difficulty in getting $\mu = \mu^2$ lies in the fact that the equalities in (1) hold for $x \in S_\mu$ only.]

To prove (iii), if S_μ is right-cancellative, for $B \subset S_\mu$, $x \in S_\mu$,

$$\begin{aligned}\mu(B) &= \mu(Bxx^{-1}) \quad [Bxx^{-1} \cap S_\mu = B]\\ &= \mu^2(Bxx^{-1}) = \mu^2(B),\end{aligned}$$

using (2). Similar is the case, when S_μ is left-cancellative.

To prove (iv), let $\varepsilon > 0$. Let K be a compact set $\subset S$. Let V be open $\supset K$ be such that $\mu(V) \leq \mu(K) + \varepsilon$ and $\mu^2(V) \leq \mu^2(K) + \varepsilon$. Now the mapping $y \to \mu^2(Ky^{-1})$ is upper semi-continuous and the mapping $y \to \mu(Vy^{-1})$ is lower semi-continuous. Therefore, for $x \in S_\mu$, there exists an open set $N(x)$ containing x such that for $y \in N(x)$,

$$\begin{aligned}\mu(Vy^{-1}) &> \mu(Vx^{-1}) - \varepsilon\\ &= \mu^2(Vx^{-1}) - \varepsilon \qquad [\text{by } (2)]\\ &\geq \mu^2(Kx^{-1}) - \varepsilon\\ &\geq \mu^2(Ky^{-1}) - 2\varepsilon\end{aligned}$$

Let $O = \bigcup_{x \in S_\mu} N(x)$ (where $N(x)$ is obtained in the above manner). Then O is open containing S_μ and therefore, by assumption, $m^n(O) \to 1$ as $n \to \infty$. Let k be such that $m^k(O) > 1 - \varepsilon$. Since $\mu * m^k = \mu$,

$$\begin{aligned}\mu(V) &= \int \mu(Vy^{-1})\ m^k(dy) \\ &\geq \int_0 \mu^2(Ky^{-1})\ m^k(dy) - 3\varepsilon \\ &\geq \int \mu^2(Ky^{-1})\ m^k(dy) - 4\varepsilon \\ &= \mu^2(K) - 4\varepsilon,\end{aligned}$$

and therefore, $\mu(K) \geq \mu(V) - \varepsilon \geq \mu^2(K) - 5\varepsilon$. Since $\varepsilon > 0$ is arbitrary, $\mu(K) \geq \mu^2(K)$. Similarly, $\mu(K) \leq \mu^2(K)$. Hence $\mu = \mu^2$.

Finally to prove (v), let $m^k(S_\mu) > 0$ for some positive integer k. Then if we consider the infinite product measure space, $(S,P) = \prod_{i=1}^{\infty} (S_i, m^k)$ where $S_i = S$ for all i, then the co-ordinate mappings $\{X_n\}$ become independent identically distributed (with distribution m^k) S-valued random variables. Hence

$$\sum_{n=1}^{\infty} P[X_n \in S_\mu] = \sum_{n=1}^{\infty} m^k(S_\mu) = \infty.$$

By Borel-Cantelli Lemma,

$P[\bigcap_{j=1}^{\infty} \bigcup_{n=j}^{\infty} X_n \in S_\mu] = 1$. This means that given $\varepsilon > 0$, $\exists\ n_o \ni$ for $n > n_o$,

$$P[\bigcup_{j=1}^{n} X_j \in S_\mu] > 1 - \varepsilon.$$

If $(m^k)^n (S_\mu^c) > \varepsilon$, then there is a compact G_δ set $A \subset S_\mu^c$ such that $(m^k)^n(A) > \varepsilon$. But

$$\begin{aligned}(m^k)^n\ (A) &= P[X_1X_2\ldots X_n \in A] \\ &= 1 - P[X_1X_2\ldots X_n \in A^c] \\ &\leq 1 - P[X_1X_2\ldots X_n \in S_\mu] \\ &\leq 1 - P[\bigcup_{j=1}^{n} X_j \in S_\mu] \\ &< 1 - (1 - \varepsilon) = \varepsilon, \text{ a contradiction.}\end{aligned}$$

Hence there is an integer $n_1 \ni m^{n_1}(S_\mu) > 1 - \varepsilon$. Now following the proof of (iv), (v) follows.

Q.E.D.

The next two results are companion results of 4.5, 4.6 and 4.7 above and indicate with more completeness what can be expected in the non-compact situation.

4.7A Proposition. Let S be a locally compact second countable semigroup generated by the support of m in P(S) and satisfying the condition:

(*) for each compact set $K \subset S$ and $x \in S$, the set Kx^{-1} is compact.

Then the following are true:

(i) for $\mu \in P(S)$, $m * \mu = \mu$ implies $\mu = \mu^2$;

(ii) for $\mu \in P(S)$, $m * \mu = \mu * m = \mu$ iff $(\frac{1}{n}) \sum_{k=1}^{n} m^k$ converge to μ weakly.

Before we present the proof of this proposition, let us mention that the condition (*) is necessary for part (ii) of this proposition. The reason is: if $S = [0, \infty)$, the non-negative reals with usual topology and multiplication, and if m = the normalized Lebesgue measure on [0, e] where $\ell ne = 1$, then it can be proved (see part A of Section 5) that $(\frac{1}{n}) \sum_{k=1}^{n} m^k$ converge to $\frac{1}{2} \cdot \delta_{\{0\}}$ in the weak-star topology, even though $m * \delta_{\{0\}} = \delta_{\{0\}} * m = \delta_{\{0\}}$.

Proof of the proposition. The 'if' part of (ii) follows easily and will not be proved. First we assume: $m * \mu = \mu$. Let us write:

$$m_n = (\frac{1}{n}) \sum_{k=1}^{n} m^k.$$

We claim that every weak*-cluster point of the sequence (m_n) is in P(S). To see this, let $m_{n_i} \to Q$ vaguely as $i \to \infty$. Let f be any continuous function with compact support. Then for any $x \in S$, the function $f_x(y) = f(yx)$ has also compact support by condition (*). Hence as $i \to \infty$, for every $x \in S$, we have:

$$g_i(x) = \int f(yx)\, m_{n_i}(dy)$$
$$\to g(x) = \int f(yx)\, Q(dy).$$

By the dominated convergence theorem, we have:

$$\int f(x)\, \mu(dx) = \int f(x)\, m * \mu(dx)$$
$$= \int f(x)\, m_{n_i} * \mu(dx)$$
$$= \int\int f(yx)\, m_{n_i}(dy)\, \mu(dx)$$
$$= \int g_i(x)\, \mu(dx)$$
$$\to \int g(x)\, \mu(dx) = \int\int f(yx)\, Q(dy)\, \mu(dx)$$
$$= \int f(x)\, Q * \mu(dx).$$

This means that $\mu = Q * \mu$ or $Q \in P(S)$. Now it is routine to check (by using continuity of convolution as $P(S) \times P(S) \to P(S)$) that $m_n \to Q$ as $n \to \infty$, $Q = Q^2$ and also $m * Q = Q * m = Q$. Then Then it is clear that S_Q, the support of Q, is the kernel of S. To see that $S_\mu \subset S_Q$, let W be any open set containing S_Q. Since $m_n * \mu = \mu$ and $m_n \to Q$ as $n \to \infty$, we have:

$$\mu(W) = \int m_n(Wx^{-1})\, \mu(dx).$$

Since for every $x \in S$, $S_Q \subset Wx^{-1}$, an open set, $m_n(Wx^{-1}) \to 1$ as $n \to \infty$ pointwise and therefore, by the dominated convergence theorem, it follows that $\mu(W) = 1$. Consequently, $S_\mu \subset S_Q$. Now we have by Proposition 3.14, for $B \subset S_Q$ and x, y in S_Q, $Q(Bx^{-1}y^{-1}) = Q(Bx^{-1})$. For $B \subset S_\mu$ and $y \in S_Q$, we have:

$$\mu(By^{-1}) = \int Q(By^{-1}z^{-1})\, \mu(dz)$$
$$= Q(By^{-1})$$

and therefore,

$$\mu(B) = \int Q(B\bar{y}^1)\ \mu(dy)$$
$$= \int \mu(B\bar{y}^1)\ \mu(dy)$$
$$= \mu^2(B).$$

Thus $\mu = \mu^2$ and the part (i) follows.

To prove (ii), we notice that if $m * \mu = \mu * m = \mu$, then since $m_n \to Q$ as $n \to \infty$, we have also:

$$\mu * Q = Q * \mu = \mu.$$

Hence $S_\mu = S_Q$, since S_μ is an ideal of S_Q which is simple. Then for any $B \subset S_\mu$,

$$\mu(B) = \int \mu(B\bar{y}^1)\ Q(dy)$$
$$= \int Q(B\bar{y}^1)\ Q(dy)$$
$$= Q^2(B) = Q(B).$$

Thus $\mu = Q$ and (ii) follows. Q.E.D.

4.7B <u>Proposition</u>. Let S be a locally compact non-compact Hausdorff semigroup generated by the support of $m \in P(S)$. Suppose S has the following condition:

(**) for each compact set $K \subset S$ and $x \in S$, the sets $K\bar{x}^1$ and $\bar{x}^1K$ are compact.

Then $m_n = (\frac{1}{n}) \sum_{k=1}^{n} m^k \to 0$ as $n \to \infty$ in the weak*-topology.

Before we prove this proposition, we remark that the condition (**) is necessary for this result. The completely simple semigroup $E \times G \times F$ has condition (**) iff E and F are both compact. It is clear that there are non-compact E and F and a compact G such that $E \times G \times F$ can support an idempotent probability measure.

<u>Proof of the proposition</u>. For simplicity, we'll assume second countability. [To get around this difficulty one can consider the one-point compactification of S, which can be made a compact semitopological semigroup in a natural way.] Let μ be a weak*-

cluster point of m_n and $\mu \neq 0$. Then using condition (**), as in the proof of Prop. 4.7A, we easily have: $\mu = \mu * m = m * \mu$. Then if we define the probability measure β by $\beta(B) = \mu(B)/\mu(S)$, we have:

$$\beta = \beta * m = m * \beta.$$

It follows from Prop. 4.7A that $\beta = \beta^2$. Therefore, by Prop. 3.15, S_β is a completely simple closed subsemigroup with its group factor compact. Suppose

$$S_\beta = E \times G \times F, \ K \text{ compact} \subset S_\beta.$$

Then for $x \in S_\beta$, $Kx^{-1} \cap S_\beta$ and $x^{-1}K \cap S_\beta$ are both compact by condition (**). This means that E and F are both compact, and therefore, S_β is compact. But S_β is an ideal of S and for $x \in S_\beta$, $S_\beta x^{-1} = S$. This means that S is compact by (**). This contradicts that S is non-compact. Hence $\mu = 0$.

Q.E.D.

In a compact Hausdorff semigroup S, let s be an arbitrary element. Let $E(s) = \{s^n : n = 1,2,\ldots\}$. Let $S(s) = \overline{E(s)}$ and G(s) be the set of all limit points of S(s). Then S(s) is a commutative semigroup and $G(s) = \bigcap_{m=1}^{\infty} \overline{\{s^n; n \geq m\}}$. Then G(s) is a group. To prove this, it suffices to show that $y.G(s) = G(s) \ \forall \ y \in G(s)$, since G(s) is commutative. If $z \in G(s)$ and $z \notin y.G(s)$, then by the compactness of G(s), we can find open set V(z) containing z, open set V(y) containing y such that

$$\overline{V(z)} \cap V(y).O = \emptyset,$$

where O is an open set containing G(s). We can find positive integers $1 \leq p < r_1 < r_2 < r_3 < \ldots$ such that $s^p \in V(y)$ and $s^{r_j} \in V(z)$, $1 \leq j < \infty$. The sequence $s^{r_j - p}$ has a limit point x in G(s). This means that $s^p x \in \overline{V(z)}$, a contradiction. Hence $G(s)$ is a group. Also $G(s)$ is an ideal (clearly) of $S(s)$ so that $G(s)$ is the kernel of $S(s)$.

We are now in a position to present a limit theorem on the convergence of $\{\mu^n\}$, $\mu \in P(S)$, S a compact group.

4.8 Theorem. Let S be a compact group and $\mu \in P(S)$. Let $\lambda = \lambda^2$ be the unit element of $G(\mu)$, and F be the minimal closed subgroup containing S_μ. The following conditions are equivalent:

(i) $\{\mu^n\}$ is convergent;

(ii) the set $\underline{\lim}\ S_{\mu^n}$ is not empty;

(iii) $\underline{\lim}\ S_{\mu^n} = \overline{\lim}\ S_{\mu^n}$;

(iv) the minimal closed subgroup containing the set $\bigcup_{n=1}^{\infty} (S_\mu)^n (S_\mu)^{-n}$ coincides with F;

(v) S_μ is not contained in any proper coset of any closed normal subgroup of F;

(vi) S_μ is not contained in any proper coset of S_λ in F;

(vii) $\lambda(Bx^{-1}) = \lambda(x^{-1}B) = \lambda(B)\ \forall\ x \in F,\ B \subset F$.

To prove this theorem, we need two lemmas.

4.9 Lemma. Let G_1 be a subgroup of $P(S)$ where S is a compact group and λ the identity of G_1. Let $S(G_1)$ be the support of G_1, i.e. $S(G_1) = \overline{\bigcup_{\mu \in G_1} S_\mu}$. Then $S(G_1)$ is a closed subgroup of S and S_λ is a normal subgroup of $S(G_1)$. Furthermore, $\mu = \lambda * \delta_g = \delta_g * \lambda$ for $\mu \in G_1$ and $g \in S_\mu$, where δ_g is the point mass at g.

Proof. We first show that $S(G_1)$ is a group. If $g_1 \in S_{\mu_1}$, $g_2 \in S_{\mu_2}$, μ_1 and $\mu_2 \in G_1$, then $g_1 g_2 \in S_{\mu_1} S_{\mu_2} \subset S_{\mu_1 * \mu_2}$,

$\mu_1 * \mu_2 \in G_1$. Also if $g \in S_\mu$, $\mu \in G_1$, then if $h \in S_{\mu^{-1}}$, we have $g.h \in S_\mu S_{\mu^{-1}} \subset S_\lambda$ and therefore $g^{-1} \in h.S_\lambda \subset S_{\mu^{-1}} S_\lambda \subset S_{\mu^{-1}}$. This means that $\bigcup_{\mu \in G_1} S_\mu$ is a group and hence its closure $S(G_1)$ is also a group.

Now if $g \in S_\mu$, $g^{-1} \in S_{\mu^{-1}}$ and therefore

$$g^{-1} S_\lambda g \subset S_{\mu^{-1}} S_\lambda S_\mu \subset S_{\mu^{-1}*\lambda*\mu} = S_\lambda$$

Since the set of all such g is dense in $S(G_1)$, it follows that S_λ is a closed normal subgroup of $S(G_1)$.

Finally let $\mu \in G_1$. Since $\mu * \lambda = \mu$, $S_\mu S_\lambda = S_\mu$ and therefore, S_μ is the union of cosets of S_λ in $S(G_1)$. If $g_1 \in S_\mu$, $g_2 \in S_\mu$, then $g_1^{-1} g_2 \in (S_\mu)^{-1} S_\mu \subset S_{\mu^{-1}} S_\mu = S_\lambda$. Hence S_μ is itself a coset of S_λ in $S(G_1)$. It follows that if $g \in S_\mu$, then S_λ is the support of $\mu * \delta_{g^{-1}}$ and $\delta_{g^{-1}} * \mu$. Also if $h \in S_\lambda$, then $\delta_h * \mu = \delta_h * (\lambda * \mu) = (\delta_h * \lambda) * \mu = \lambda * \mu = \mu$ and hence $\delta_h * \mu * \delta_{g^{-1}} = \mu * \delta_{g^{-1}}$. By the uniqueness of Haar measure, $\mu * \delta_{g^{-1}} = \lambda$ or $\mu = \lambda * \delta_g$. Similarly,

$$\mu = \delta_g * \lambda \ \forall g \in S_\mu.$$

Q.E.D.

4.10 <u>Corollary</u>. Let $\mu \in P(S)$ and λ be the identity of $G(\mu)$, S a compact Hausdorff group. Then S_μ is contained in a certain coset of the normal subgroup S_λ of the group $S(G(\mu))$ and therefore, is contained in a certain two-sided coset of the subgroup S_λ of the group S.

<u>Proof</u>. Since $G(\mu)$ is the kernel of $S(\mu)$, $S(\mu) * \lambda = G(\mu)$ or $\mu * \lambda \in G(\mu)$. Hence, by lemma 4.9, $\mu * \lambda = \lambda * \delta_g = \delta_g * \lambda$ or $S_\mu S_\lambda \subset S_\lambda g$ or $S_\mu \subset S_\lambda g = gS_\lambda$ where $g \in S(G(\mu))$. Q.E.D.

4.11 Lemma. Let $\mu \in P(S)$, S a compact group. Then

(i) $S(G(\mu)) = \overline{\lim}\, S_{\mu^n} = S(S(\mu)) = F$,

where F is the minimal closed group containing S_μ ;

(ii) If $\underline{\lim}\, S_{\mu^n} \neq \phi$, then

$$S(G(\mu)) = \underline{\lim}\, S_{\mu^n}.$$

Proof. Now $\overline{\lim}\, S_{\mu^n} = \bigcap_{m=1}^{\infty} \overline{\bigcup_{n=m}^{\infty} S_{\mu^n}}$ and

$G(\mu) = \bigcap_{m=1}^{\infty} \overline{\{\mu^n : n \geq m\}}$. It is clear that

$S(G(\mu)) \subset \overline{\lim}\, S_{\mu^n} \subset S(S(\mu)) \subset F$. Since by Lemma 4.9, $S(G(\mu))$ is a closed subgroup and since $S_\mu \subset S(G(\mu))$, (i) follows.

To prove (ii), we show first that $\underline{\lim}\, S_{\mu^n}$, when non-empty, is an ideal of $S(G(\mu))$.

Now $\underline{\lim}\, S_{\mu^n} = \overline{\bigcup_{m'=1}^{\infty} \bigcap_{n=m'}^{\infty} S_{\mu^n}}$. Let $g_1 \in S_{\mu^m}$, $g_2 \in \underline{\lim}\, S_{\mu^n}$. Let U be a neighborhood of $g_1 g_2$. Then $\exists$ a neighborhood V of g_2 such that $g_1 V \subset U$. Now there is a $m_o \ni \forall\, n > m_o$, $V \cap S_{\mu^n} \neq \phi$. Let $n > m + m_o$ and $g_3 \in V \cap S_{\mu^{n-m}}$. Then $g_1 g_3 \in g_1 V \subset U$ and also $g_1 g_3 \in S_{\mu^m} S_{\mu^{n-m}} \subset S_{\mu^n}$. Hence $U \cap S_{\mu^n} \neq \phi\ \forall\, n > m + m_o$. Hence $g_1 g_2 \in \underline{\lim}\, S_{\mu^n}$. Similarly, $g_2 g_1 \in \underline{\lim}\, S_{\mu^n}$. This means that $\underline{\lim}\, S_{\mu^n}$ is an ideal of $S(S(\mu)) = S(G(\mu))$. The rest follows easily.

Q.E.D.

We now prove Theorem 4.8.

Proof of Theorem 4.8. Suppose $\{\mu^n\}$ is convergent to ν.

Let $g \in S_\upsilon$. If $\underline{\lim}\, S_{\mu^n} = \phi$, then there is a neighborhood U of g such that $U \cap S_{\mu^{n_i}} = \phi$, for some subsequence $\{n_i\}$ of natural numbers. Now

$$g \in U \cap S_\upsilon \subset U \cap \overline{S(\{\mu^{n_i}, i = 1,2,\ldots\})}.$$

Also since $U \cap S_{\mu^{n_i}} = \phi \;\forall\, i$,

$$S_{\mu^{n_i}} \subset U^c \text{ or } \overline{S(\{\mu^{n_i}, i = 1,2,\ldots\})} \subset U^c,$$ which is a contradiction. This proves that $S_\upsilon \subset \underline{\lim}\, S_{\mu^n}$ and therefore (i) => (ii). (ii) => (iii) by Lemma 4.11. To prove (iii) => (iv), we show that

$$\underline{\lim}\, S_{\mu^n} \subset \overline{\bigcup_{n=1}^{\infty} S_{\mu^n} \overline{(S_{\mu^n})^{-1}}}.$$

To show this, let $y \in \underline{\lim}\, S_{\mu^n}$. Then since $\underline{\lim}\, S_{\mu^n}$ is a group, $y = z.w^{-1}$, where z and w are in $\underline{\lim}\, S_{\mu^n}$. Given any neighborhood V of y, there exist neighborhoods V_1 of z and V_2 of w such that $V_1 V_2^{-1} \subset V$. Clearly there exist n such that

$$V_1 \cap S_{\mu^n} \neq \phi,\; V_2 \cap S_{\mu^n} \neq \phi$$

so that $V \cap S_{\mu^n} (S_{\mu^n})^{-1} \neq \phi$. Hence the above inclusion. Now we show that (iv) => (v). Suppose $S_\mu \subset H.g$, where H is a proper normal subgroup of F and $g \in F$. Then $S_\mu^n \subset H.g^n$ and $S_\mu^{-n} \subset g^{-n}H$, so that $S_\mu^n S_\mu^{-n} \subset H$ which implies $F = H$, a contradiction. Therefore (iv) => (v). The implication (v) => (vi) is now obvious, since S_λ is a normal subgroup of F. To prove (vi) => (vii), we notice from Corollary 4.10 that $S_\mu \subset S_\lambda.g$ for some $g \in F$. This means that $S_\lambda.g = F$ or

$S_\lambda = F$. Hence (vii) is obvious, since λ being idempotent, is the Haar measure on its support. To prove (vii) => (i), let $\lambda_1 \in G(\mu)$. Then $\lambda_1 * \lambda = \lambda_1$, since λ is the identity of the group $G(\mu)$. Also since λ is the Haar measure of $F \supset S(G(\mu))$, $\lambda_1 * \lambda = \lambda$. Hence $G(\mu) = \{\lambda\}$. Since $G(\mu)$ consists of the limit pts. of $\{\mu^n : n \geq 1\}$, μ^n converges to λ.

Q.E.D.

4.12 Theorem. Let $\mu \in P(S)$, S a compact group and let $\lambda = \lambda^2 \in G(\mu)$ be the unit element in $G(\mu)$. Let $S_\lambda . g = g . S_\lambda$ be the two-sided coset of S_λ in S to which S_μ belongs. Then if $h \in S_\lambda . g$, the sequence $\delta_{h^{-n}} * \mu^n = 1,2,\ldots$ converges to λ.

Proof. Let υ be a limit-point of $\{\delta_{h^{-n}} * \mu^n\}$. Then $\upsilon = \delta_{g_1} * \upsilon'$, where g_1 is a limit-point of $\{h^{-n}\}$ and υ', a limit-point of $\{\mu^n\}$. Since $\upsilon' \in G(\mu)$, $\upsilon' = \delta_{g_2} * \lambda$ or $\upsilon = \delta_{g_1 g_2} * \lambda$. Since the support of $\delta_{h^{-n}} * \mu^n$ $(\forall n)$ is contained in S_λ, $S_\upsilon \subset S_\lambda$. Hence since λ is the Haar measure on S_λ, $\delta_{g_1 g_2} * \lambda = \lambda$. The theorem follows.

Q.E.D.

4.13 Theorem. Let μ be a regular probability measure on a compact Hausdorff semigroup S which is generated by S_μ. The sequence $\{\mu^n\}$, $n = 1,2,\ldots$ will not converge as $n \to \infty$ iff there is a proper closed normal subgroup G' of G such that $YX \subset G'$ and

$$S_\mu (X \times G' \times Y) \subset X \times gG' \times Y$$

where $g \notin G'$, $X \times G \times Y$ being the standard representation of the completely simple kernel K of S.

Proof. (The 'if' part). Suppose that G' is a proper closed normal subgroup of G' with Y X $\subset$ G' such that

$$S_\mu (X \times G' \times Y) \subset X \times gG' \times Y$$

where $g \notin G'$. Then

$$S_\mu^2 (X \times G' \times Y) \subset S_\mu (X \times gG' \times Y)$$

$$= [S_\mu (X \times G' \times Y)] (X \times gG' \times Y)$$

$$\subset X \times g^2 G' \times Y$$

By induction, it follows that for each positive integer n,

$$S_\mu^n (X \times G' \times Y) \subset \overline{X \times g^n G' \times Y}$$

and consequently, $G = \overline{\bigcup_{n=1}^{\infty} g^n G'}$ since $S = \overline{\bigcup_{n=1}^{\infty} S_\mu^n}$. Let β be a probability measure with support in $X \times G' \times Y$. Then the support of $\mu^n * \beta$ is contained in $X \times g^n G' \times Y$.

If $\{g^n G' \mid n \geq 1\}$ is finite, then G' and gG' are both open sets; and, for infinitely many n,

the support of $\mu^n * \beta \subset X \times G' \times Y$

and for infinitely many other values of n,

the support of $\mu^n * \beta \subset X \times gG' \times Y$.

Clearly then, the sequence (μ^n) cannot converge weakly. Now if $\{g^n G' \mid n \geq 1\}$ is infinite, then, the sequence g^n has a cluster point h and consequently, hg $(\neq h)$ is another cluster point of this sequence. Since $hG' \cap hgG' = \phi$, there are disjoint open sets V_1 and V_2 such that $hG' \subset V_1$ and $hgG' \subset V_2$. It is clear now that for infinitely many n,

the support of $\mu^n * \beta \subset X \times V_1 \times Y$

and for infinitely many other values of n,

the support of $\mu^n * \beta \subset X \times V_2 \times Y$.

Hence, the sequence $\mu^n * \beta$ cannot converge weakly and consequently, the sequence μ^n is not weakly convergent.

(The 'only if' part). Let us now assume that the sequence μ^n is not weakly convergent. Let $K(\mu)$ be the kernel of the compact commutative semigroup $\overline{E(\mu)} = \{\mu^n \mid n \geq 1\}^-$. Then $K(\mu)$ is a group and $K(\mu)$ consists of all limit points (including subsequential limits) of $\overline{E(\mu)}$. Since μ^n does not converge weakly, there is an element $\eta_1 \in K(\mu)$, different from η, the identity of $K(\mu)$. Now by Prop. 4.4, $S_\eta \subset K = X \times G \times Y$. Since $\mu * \eta \in K(\mu)$, there is $\eta' \in K(\mu)$ such that,

$$\mu * \eta * \eta' = \eta' * \eta * \mu = \eta.$$

Since S_μ generates S, this means that $S\eta$, being completely simple, must be $X \times G_1 \times Y$, where G_1 is a compact subgroup of G. By Theorem 3.16, $\eta = \beta_1 * W_{G_1} * \beta_2$ where $\beta_1 \in P(X)$, $\beta_2 \in P(Y)$ and W_{G_1} is the normed Haar measure on G_1. The set G_1 must be actually a proper subgroup of G since otherwise

$$\begin{aligned}\eta_1 &= \eta * \eta_1 * \eta \\ &= \beta_1 * W_{G_1} * (\beta_2 * \eta_1 * \beta_1) * W_{G_1} * \beta_2 \\ &= \beta_1 * W_{G_1} * W_{G_1} * W_{G_1} * \beta_2 = \eta,\end{aligned}$$

a contradiction. Note that $\beta_2 * \eta_1 * \beta_1 \in P(G)$. Now $\mu * \eta = \eta$ implies that $\mu^n * \eta = \eta$ and this means that $\eta_1 * \eta = \eta$, a contradiction. Hence $\mu * \eta \neq \eta$. Since $(\mu * \eta) * \eta = \eta * (\mu * \eta) = \mu * \eta$, it is clear that $S_{\mu*\eta} = X \times CG_1 \times Y = X \times G_1C \times Y$, where C is a closed subset of elements of G. Since there exists $\eta' \in K(\mu)$ such that $(\mu * \eta) * \eta' = \eta' * (\mu * \eta) = \eta$, it follows that $S_{\mu*\eta} = X \times gG_1 \times Y$, where $gG_1 = G_1g$. Here $g \notin G_1$, since otherwise

$$\begin{aligned}\mu * \eta &= \eta * (\mu * \eta) * \eta \\ &= \eta, \text{ as before.}\end{aligned}$$

Since $S_\mu\ (X \times G_1 \times Y) = X \times gG_1 \times Y$, it follows that $S_\mu^n\ (X \times G_1 \times Y) = X \times g^nG_1 \times Y$ and therefore, since S_μ generates S, $G = \overline{\bigcup_{n=1}^{\infty} g^nG_1}$. Since $gG_1 = G_1g$, it is now clear that G, is a normal subgroup of G.

Q.E.D.

Now we consider the weak-star convergence (or vague convergence) in a completely simple semigroup.

4.14 <u>Theorem.</u> Let S be a locally compact second countable semigroup completely simple with usual product representation $X \times G \times Y$. Suppose $\mu \in P(S)$ and $S = \bigcup_{n=1}^{\infty} F^n$, where F is the support of μ. Then $\mu^n \to 0$ vaguely as $n \to \infty$ iff the group factor G is non-compact.

<u>Proof.</u> First, we notice that given $\varepsilon > 0$, there exists a compact set $K \subset S$ such that $\mu(K) > 1 - \varepsilon$. Therefore we can find compact subsets $K_1 \subset X$, $K_3 \subset Y$ such that $K \subset K_1 \times G \times K_3$ and

(3) $\mu(K_1 \times G \times K_3) > 1 - \varepsilon.$

Then we have:

$$\mu^2(K_1 \times G \times Y) = \int \mu((K_1 \times G \times Y)z^{-1})\mu(dz)$$
$$\geq \int \mu(K_1 \times G \times Y)\mu(dz)$$
$$\geq 1 - \varepsilon, \text{ by (3)}$$

and similarly, for every $n \geq 1$

$$\mu^n(K_1 \times G \times Y) \geq 1 - \varepsilon \text{ as well as}$$
$$\mu^n(X \times G \times K_3) \geq 1 - \varepsilon.$$

Hence we have:

(4) $\mu^n(K_1 \times G \times K_3) \geq 1 - 2\varepsilon.$

This means that if G is compact, the sequence μ^n has cluster points only in $P(S)$. This proves the "only if" part of the theorem.

To prove the "if part" of the theorem we assume that G

is non-compact and the sequence μ^n do not converge vaguely to zero. We will prove the theorem by reaching a contradiction to this assumption. First, we need to observe the following:

(5) For compact subsets A, B of S and a compact subset $K \subset X$, the set $A^{-1}B \cap (K \times G \times Y)$ is compact.

To prove (5), it can be easily verified that $A^{-1}B$ is a closed subset of S. We may and do assume that $A \subset A_1 \times A_2 \times A_3$ and $B \subset B_1 \times B_2 \times B_3$, where A_1, B_1 are compact subsets of X while A_2, B_2 are compact subsets of G and A_3, B_3 are compact subsets of Y. Then one can verify that

$(A_1 \times A_2 \times A_3)^{-1}(B_1 \times B_2 \times B_3) \cap K \times G \times Y \subset K \times [(A_3K)^{-1}A_2^{-1}B_2] \times B_3$,

which is compact noticing that $A_3.K$ is a compact subset of G. Now the assertion in (5) is clear.

We will now divide the proof into four steps.

<u>Step 1</u>. Here we will prove a lemma similar to that given by Csiszar [13] in the case when S is a group. His arguments have to be modified somewhat. We claim:

Let $b(K) = \lim_{n\to\infty} a_n(K)$, where

$a_n(K) = \sup\{\mu^n(K^{-1}) : x \in S\}$,

and K is a compact set whence

(6) $a_n(K) \geq a_{n+1}(K)$. If

$b = \sup\{b(K) : K \text{ compact} \subset S\}$,

then either $b = 0$ or $b = 1$.

To prove this claim, suppose $0 < b < 1$. Then we choose c such that $b < c < 1$ and $0 < c. \frac{1+c}{2} < b$. Let D be an <u>arbitrary</u> compact set $\subset S$. Then there exists a positive integer k such that

(7) $\sup_{x\in S} \mu^k(Dx^{-1}) < c.$

Let $\varepsilon > 0$ be such that $\varepsilon < \frac{c}{2}$ and

(8) $\sup_{x\in S} \mu^k(Dx^{-1}) < c - \varepsilon$

and A be a compact set such that

(9) $\mu^k(S-A) < \frac{c}{2} - \varepsilon$

Also by (4), we can find compact $K_1 \subset X$ such that for all $n \geq 1$,

(10) $\mu^n(K_1 \times G \times Y) > 1 - \varepsilon.$

Let $E = A^{-1}D \cap (K_1 \times G \times Y)$. Then by (5), E is compact. Since for y in $K_1 \times G \times Y$,

$$Dx^{-1}y^{-1} \cap A \neq \phi \Rightarrow y\, x \in E \Rightarrow y \in Ex^{-1},$$

we have by (9),

$$y \in (K_1 \times G \times Y) \cap (S - Ex^{-1}) \text{ implies}$$

(11) $\mu^k(Dx^{-1}y^{-1}) < \frac{c}{2} - \varepsilon$

Therefore for $n > k$, we have

$$\mu^n(Dx^{-1}) = \int \mu^k(Dx^{-1}y^{-1})\mu^{n-k}(dy)$$

$$\leq \int_{Ex^{-1}} + \int_{(S-Ex^{-1}) \cap (K_1xGxY)} + \int_{S-(K_1xGxY)}$$

$$< (c-\varepsilon)\mu^{n-k}(Ex^{-1})$$

$$+ (\frac{c}{2} - \varepsilon)[1 - \mu^{n-k}(Ex^{-1})] + \varepsilon$$

$$= \frac{c}{2} + \frac{c}{2}\mu^{n-k}(Ex^{-1})$$

for all $x \in S$. We also observe that there exists a positive integer N such that $n > N$ implies:

(12) $\mu^{n-k}(Ex^{-1}) < c$ for every x.

By (12) and (13), $n > N \Rightarrow$ for every $x \in S$,

$$\mu^n(Dx^{-1}) \leq \frac{c}{2} + \frac{c^2}{2} = \frac{c(1+c)}{2} < b.$$

This means that $b(D) \leq \frac{c(1+c)}{2}$ for any compact set D so that

$b = \sup\{b(D) : D \text{ compact}\} < b$, a contradiction. This proves (6) and Step I is complete.

Step II. In this step, we will show that the set $\overline{\{\mu^n : n \geq 1\}} \cap P(S)$ is non-empty.

Since we have assumed that the sequence μ^n doesn't converge to zero vaguely as $n \to \infty$, there exists $\delta > 0$ such that for some subsequence (n_k) of positive integers, $\mu^{n_k}(K) > \delta > 0$ for some compact set K. By Step I, the real number b (defined there) is now 1 since

$$\delta < \mu^{n_k}(K) = \int \mu^{n_k - 1}(Ky^{-1})\mu(dy)$$

$$\leq \sup_{y \in S} \mu^{n_k - 1}(Ky^{-1}) .$$

This means that given ε such that $0 < \varepsilon < \delta$, there is a compact set B and elements $x_n \in S$ such that for all $n \geq 1$,

(14) $\mu^n(Bx_n^{-1}) > 1 - \varepsilon > 1 - \delta.$

Hence $Bx_{n_k}^{-1} \cap K \neq \phi$ for all k and therefore, the x_{n_k}'s all belong to $K^{-1}B$. By (4), we can find compact sets $K_1 \subset X$, $K_3 \subset Y$ such that for all $n \geq 1$,

$$\mu^n(K_1 \times G \times K_3) > 1 - \varepsilon .$$

Since $Bx_{n_k}^{-1} \subset B(K^{-1}B)^{-1}$, we have for all k,

$$\mu^{n_k}(B(K^{-1}B)^{-1} \cap (K_1 \times G \times K_3)) > 1 - 2\varepsilon.$$

Writing $C = K^{-1}B \cap (K_1 \times G \times Y)$, we see that C is compact and it can be easily verified that for all k,

$$\mu^{n_k}(BC^{-1} \cap (K_1 \times G \times K_3)) > 1 - 3\varepsilon.$$

This means that given $\varepsilon > 0$, we have found a compact set M

such that

$$\mu^{n_k}(M) > 1 - 3\varepsilon \text{ for all } k.$$

Hence the cluster points of (μ^{n_k}) are all in $P(S)$ and this proves Step II.

Step III. In this step, we will show that there is a cluster point of (μ^n) which is an idempotent probability measure. We will employ Csiszar's method of tail idempotents, [13].

By Step II, there is a subsequence (n_i) of positive integers such that $\mu^{n_i} \to \beta \in P(S)$ vaguely as $i \to \infty$. Since the sequence

$$(\mu^{n_i}, \mu^{n_i - 1}, \ldots, \mu^2, \mu, 0, 0, \ldots)$$

are elements in the compact space $\overset{\infty}{\underset{j=0}{X}} B(S)_j$, $B(S)_j = B(S)$ for all j, it follows that there exists a subsequence $(p_i) \subset (n_i)$ such that for each non-negative integer k,

(15) $\mu^{p_i - k} \to \mu_k \in B(S)$ vaguely as $i \to \infty$. Since the convolution in $B(S)$ is not even separately continuous for a general locally compact semigroup, the fact that

(16) $\mu^k * \mu_k = \beta$, $0 < k < \infty$

needs some justification.

For this, let $f \in C(S)$ and $\varepsilon > 0$. By (4), we can find compact sets $K_1 \subset X$, $K_3 \subset Y$ such that for each positive integer n,

(17) $\mu^n(K_1 \times G \times K_3) > 1 - \varepsilon$.

This means that

(18) $\mu_k((X - K_1) \times G \times Y) < \varepsilon$

Let K_ε be any compact set with

(19) $\mu^k(K_\varepsilon) > 1 - \dfrac{\varepsilon}{2||f||}$

and let us define:

(20) $g(y) = \int_{K_\varepsilon} f(xy)\mu^k(dx)$.

Then $g(y) = 0$ if $y \notin K_\varepsilon^{-1}K_o$, where k_o = the compact support of f. Let U, V be open sets such that $K_1 \subset U$, $K_3 \subset V$ and $\bar{U}$, $\bar{V}$ are compact. There exists a continuous function h from S into [0,1] such that

$$h(y) = 0, \; y \notin U \times G \times V$$

(21) $$= 1, \; y \in K_1 \times G \times K_3$$

Then g(y).h(y) is continuous with compact support $K_\varepsilon^{-1}K_o \cap \bar{U} \times G \times \bar{V}$. Hence

$$\int g(y)h(y)\mu^{p_i-k}(dy) \to \int g(y)h(y)\mu_k(dy)$$

as $i \to \infty$. Since $g(y)h(y) = g(y)$ for $y \in K_1 \times G \times K_3$, by (17) and (18) there exists i_o such that $i \geq i_o$ implies

(22) $$\left|\int g(y)\mu^{p_i-k}(dy) - \int g(y)\mu_k(dy)\right| < 2\varepsilon.$$

By (17) and (18) again, we have for $i \geq i_o$

$$\left|\iint f(xy)\mu^k(dx)\mu^{p_i-k}(dy) - \iint f(xy)\mu^k(dx)\mu_k(dy)\right| < 4\varepsilon.$$

This proves (16).

Since $\beta \in P(S)$, it follows from (16) that $\mu_k \in P(S)$. Let $(q_i) \subset (p_i)$ be a subsequence such that

(23) $\mu_{q_i} \to Q \in B(S)$, as $i \to \infty$.

Then for $k < j < q_i$,

$$\mu^{q_i-k} = \mu^{j-k} * \mu^{q_i-j}$$

Using the same argument as used to establish (16), we have

(24) $\mu_k = \mu^{j-k} * \mu_j$, $k < j < \infty$

Writing $j = q_i$, we have:

$$\mu_k = \mu^{q_i - k} * \mu_{q_i}$$

Again arguing as in (16), we have

(25) $\mu_k = \mu_k * Q$

which means that $Q \in P(S)$ and by (23) and (25),

$$Q = Q * Q.$$

Since S is a separable metric space, P(S) is at least first countable and therefore by (15) and (23), we can find a subsequence $(r_i) \subset (q_i)$ such that

(26) $\mu^{r_{i+1} - r_i} \to Q$ as $i \to \infty$.

This completes step III.

<u>Step IV</u>. In this step, we will complete the proof of the theorem.

First, we observe that in the beginning of Step III, we could assume that β was an idempotent probability measure (by replacing β by Q). Then equation (16) will read:

$$\mu_k * \mu^k = \mu^k * \mu_k = Q = Q^2 \in P(S)$$

for all positive integers k. This means that

$$\text{supp}\mu_k \cdot F^k \subset \text{supp}Q \text{ ; also}$$

$$F^k \cdot \text{supp}\mu_k \subset \text{supp}Q,$$

where $F = \text{supp}\mu$. Since $S \doteq \overline{\bigcup_{k=1}^{\infty} F^k}$, it is clear that the support of Q, which is completely simple since Q is idempotent (see [ ]), is $X \times G_1 \times Y$ where G_1 is a compact subgroup of G and $Y.X \subset G_1$.

We now claim

(28) $F \subset X \times gG_1 \times Y$

for some $g \in G - G_1$ and G_1 is a <u>normal</u> subgroup of G.

To prove (28), let (x_1,g_1,y_1) and (x_2,g_2,y_2) be elements in F. Let $(x',g',y') \in \text{supp}(\mu_1)$. Then by (25),

$$(x_1,g_1,y_1)(x',g',y') \in X \times G_1 \times Y \text{ or}$$

(29) $g_1(y_1x')g' \in G_1$ or

$$g' \in G_1 g_1^{-1} G_1.$$

Also by (25), we have

$$(x',g',y')(x_2,g_2,y_2) \in X \times G_1 \times Y \text{ or}$$

$$g'(y'x_2)g_2 \in G_1 \text{ or}$$

$$g_2 \in G_1\, g'^{-1} G_1 \text{ or}$$

(30) $g_2 \in G_1, g_1 G_1$, by (29).

This proves the following fact:

(31)
$$F \subset X \times G_1 g G_1 \times Y$$
$$\text{and } \text{supp}\mu_1 \subset X \times G_1 g^{-1} G_1 \times Y.$$

Now using the same method as used in deriving equation (25) and writing

$$\mu^{q_i-k} = \mu^{q_i-j} * \mu^{j-k}$$

for $k < j < q_i$, we get the following identity:

(32) $$\mu_k = Q * \mu_k .$$

From (32), we have:

$$(X \times G_1 \times Y).\ \text{supp}\ \mu_1 \subset \text{supp}\ \mu_1.$$

From (31), it is clear that a typical element in $\text{supp}\mu_1$ can be taken as $(x,h_1g^{-1}h_2,y)$ where h_1 and h_2 are in G_1, and therefore,

$$(X \times G_1 \times Y).(x,h_1g^{-1}h_2,y) = X \times G_1 g^{-1} h_2 \times \{y\} \subset \text{supp}\ \mu_1.$$

Now from (27), we have:

$$F.\text{supp}\mu_1 \subset X \times G_1 \times Y$$

and therefore, by choosing a typical element in F (using (31)) as (x_1, h_3gh_4, y_1), where h_3 and h_4 are in G_1, we have:

$$(x_1, h_3gh_4, y_1).(X \times G_1g^{-1}h_2 \times \{y\}) \subset X \times G_1 \times Y.$$

This means that

$$h_3.gG_1g^{-1}h_2 \subset G_1 \quad \text{or} \quad gG_1 \subset G_1g.$$

Similarly, using (25) and (31), we have: $G_1g \subset gG_1$.
Hence $gG_1 = G_1g$.
This proves that for $1 \leq k < \infty$

$$F^k \subset X \times g^kG_1 \times Y$$

and therefore,

$$G = \overline{\bigcup_{k=1}^{\infty} g^kG_1}.$$

This means that G_1 is a normal subgroup of G and (28) is proven. Now the sequence g^k $(1 \leq k < \infty)$ cannot be completely contained in any compact set since G is non-compact. This means that the sequence g^k must converge to ∞ as $k \to \infty$ (see [24], page 85).

Therefore given <u>any</u> compact set $K_2 \subset G$, there exists a positive integer N (depending upon K_2) such that for $k > N$,

$$g^k.G_1 \cap K_2 = \phi.$$

Hence for all $k > N$,

$$F^k \cap (X \times K_2 \times Y) = \phi$$

which contradicts our original assumption that the sequence μ^n don't converge to zero vaguely. The proof of the theorem is now complete.

We remark here that in the above proof second countability has been used crucially at least in deriving (26). It is not clear how one can prove the same result without the assumption of second countability. However, when S is a locally compact non-compact, but compactly generated group generated by the support of a probability measure μ, the convolution sequence μ^n converges to 0 vaguely as n tends to infinity. The reason is: by Theorem 8.7 in [24], in this case there is a compact normal subgroup H of S such that the quotient group $\frac{S}{H}$ is a locally compact non-compact second countable group and therefore, if P is the probability measure on this quotient induced by μ, then by our theorem, P^n converges to 0 vaguely and this means that μ^n also converges to 0 vaguely as n tends to infinity.

We further remark that in a locally compact non-compact connected group which is generated by the support of a probability measure μ, we conjecture the following stronger result: for every compact set K,

$$\sup\{\mu^n(Kx) : x \text{ in } S\} \to 0 \quad \text{as } n \to \infty ;$$

in the abelian case, this is easy to show by using the structure theorem for abelian groups. In the discrete situation, this result is, of course, not true; for example, take S = the integers and μ = the unit mass at 1.

We now present a result on the vague convergence of μ^n on certain other locally compact semigroups. To do this, let us say that a topological semigroup S satisfies the condition (CR) if for x, y in S, $x \notin Sy$ implies that there exist open sets V(x) and V(y) containing x and y respectively such that $V(x)V(y)^{-1}$ is empty. Condition (CL) is the left analogue of (CR) and defined accordingly. Every completely simple or discrete semigroup satisfies (CL) and (CR). First, we state an interesting result (taken from [47]) which will be proven in Chapter 2, 5.9.

4.14A Proposition. Suppose S is a locally compact semigroup satisfying (CL) and (CR), and generated by the support of μ in P(S). Suppose there exists an element x in S with the property :

$$(*) \quad \sum_{n=1}^{\infty} \mu^n(N(x)) = \infty \text{ for every open set } N(x) \text{ containing } x.$$

Then S has a completely simple kernel K which consists of only those and all those points with property (*).

We will use 4.14A to prove

4.14B Proposition. Suppose that S is a locally compact noncompact semigroup which is abelian and has condition (L): For compact sets A,B, the set AB^{-1} is compact. Suppose that S is generated by the support of μ in P(S). Then the sequence μ^n converges to 0 vaguely as n tends to infinity.

Before we prove this theorem, let us note that the interval (0,1) with multiplication and usual topology is an abelian noncompact semigroup with condition (L). Though 4.14B does not answer the question of vague convergence of μ^n on general noncompact semigroups, let us remark that the abelian assumption and condition (L) are both necessary in 4.14B. [The reader can verify this by looking at the completely simple semigroup $E\times G\times F$ with E noncompact and G, F compact, and also by looking at Example A in section 5.]

Proof of Proposition 4.14B. First, we observe that " abelian" and (L) imply (CL) and (CR). Suppose μ^n does not converge vaguely to 0 as n tends to infinity. Then by 4.14A, S has a completely simple kernel K, which is a group. By condition (L), K is noncompact. If $e=e^2$ is in K, then by 4.14, $\mu^n*\delta_e = (\mu*\delta_e)^n$ converges to 0 and this means that μ^n converges to 0 as $n \to \infty$.

Q.E.D.

In 4.13, we have seen that the sequence μ^n need not converge weakly in a compact semigroup. However, we have the following useful theorem.

4.15 Theorem. Let S be a compact second countable semigroup. Let $\mu \in P(S)$ and $S = \overline{\bigcup_{n=1}^{\infty} F^n}$, where F is the support of μ. Then there exist elements $a_n \in S$ such that the sequence $\mu^n * \delta_{a_n}$ converge weakly to some $\lambda \in P(S)$ as $n \to \infty$.

Proof. By the same kind of trick used in Step III of Theorem 4.14, there exists a subsequence n_k of positive integers such that for each non-negative integer j,

$$\mu^{n_k - j} \to \mu_j \in P(S) \text{ as } k \to \infty$$

(33)

$$\text{and } \mu_{n_k} \to \mu_\infty = \mu_\infty^2 \in P(S) \text{ as } k \to \infty$$

Since S is locally compact second countable, S is a separable metric space and therefore, every closed set is a G_δ-set (i.e., countable intersection of open sets). Let (0_n) be a decreasing sequence of open sets such that $S_{\mu_\infty} = \bigcap_{n=1}^{\infty} 0_n$.

Since $P(S)$ is in this case first countable (with respect to weak or weak*-topology), it follows from (33) that there exists a subsequence $(p_k) \subset (n_k)$ such that

$$\mu^{p_{k+1} - p_k} \to \mu_\infty \text{ as } k \to \infty \text{ and}$$

(34)

$$\mu^{p_{k+i} - p_k}(0_k) > 1 - \frac{1}{k} \text{ for each } i \geq 1.$$

Let m be a positive integer such that $p_k < m \leq p_{k+1}$. Then

$$\mu^{p_{k+2} - p_k}(0_k) = \int \mu^{m - p_k}(0_k y^{-1}) \mu^{p_{k+2} - m}(dy).$$

Hence there exists z_m such that

$$\text{(35)} \quad \mu^{m - p_k}(0_k z_m^{-1}) > 1 - \frac{1}{k}.$$

Let z be any element in S_{μ_∞}. Then we claim: the sequence

(36) $\mu^n * \delta_{z_n z} \to \mu_o * \delta_z$ weakly as $n \to \infty$.

To prove the claim, let Q be any cluster point of the sequence $\mu^n * \delta_{z_n z}$. Then there is a subsequence m_j of positive integers such that

$$\mu^{m_j} * \delta_{z_{m_j} z} \to Q \text{ weakly}$$

as $j \to \infty$. We now replace the sequence m_j by a suitable subsequence (and still calling this subsequence the m_j's) such that we can choose a subsequence (p_{k_j}) of the sequence (p_k) such that

$$p_{k_j} < m_j \leq p_{k_j+1}$$

Now the sequence $\mu^{m_j - p_{k_j}} * \delta_{z_{m_j}}$ has a cluster point of the form $\lambda * \delta_{z_o}$, where z_o is a cluster point of the z_{m_j}'s and λ is a cluster point of the sequence $\mu^{m_j - p_{m_j}}$. It is clear from the choice of the z_m's in (35) that

(37) $\operatorname{supp}(\lambda * \delta_{z_o}) \subset S_{\mu_\infty}$

Since μ_∞ is the identity of the kernel (which is a group) of $\{\mu^n : n \geq 1\}^-$ and the kernel consists of the set of all the subsequential limits of $\{\mu^n : n \geq 1\}$, we have:

(38) $\mu_\infty * \lambda * \delta_{z_o} = \lambda * \delta_{z_o}$

Now since μ_∞ is an idempotent probability measure, we have from 3.14: for any Borel set B,

$$\mu_\infty(Bz^{-1}y^{-1}) = \mu_\infty(Bz^{-1})$$

for any $y \in S_{\mu_\infty}$. [Recall $z \in S_{\mu_\infty}$]. Therefore by (37) and (38),

$$\lambda*\delta_{z_o z}(B) = \lambda*\delta_{z_o}(Bz^{-1})$$

$$= \mu_\infty*\lambda*\delta_{z_o}(Bz^{-1})$$

$$= \int \mu_\infty(Bz^{-1}y^{-1})\,\lambda*\delta_{z_o}(dy)$$

$$= \int \mu_\infty(Bz^{-1})\;\lambda*\delta_{z_o}(dy)$$

$$= \mu_\infty(Bz^{-1})$$

$$= \mu_\infty*\delta_z(B).$$

Since $\mu^{m_j}*\delta_{z_{m_j}} = \mu^{p_{k_j}}*(\mu^{m_j-p_{k_j}}*\delta_{z_{m_j}z})$, and by (33), $\mu^{p_{k_j}} \to \mu_o$ weakly as $j \to \infty$, it is clear that

$$Q = \mu_o * (\lambda * \delta_{z_o z})$$

$$= \mu_o * (\mu_\infty * \delta_z)$$

$$= \mu_o * \delta_z.$$

This proves our claim (36). The theorem now follows.

Q.E.D.

We remark that Theorem 4.15 doesn't extend to the non-compact case, even in the case of locally compact groups. For example, let S be a locally compact non-compact abelian group which is generated by the support F of some probability measure $\mu \in P(S)$ such that F contains the identity of S. [For instance, take μ to be the normalized Lebesgue measure on $[-2,2]$ and $S = (-\infty,\infty)$.] Now, if there exist $a_n \in S$ such that $\mu^n * \delta_{a_n} \to \lambda \in P(S)$ weakly as $n \to \infty$, then $\mu^n * \delta_{a_n} * \delta_{a_n^{-1}} * \bar{\mu}^n = (\mu * \bar{\mu})^n$ converge to $\lambda*\bar{\lambda}$ as $n \to \infty$, where for any $\beta \in P(S)$, $\bar{\beta}$ is defined by $\bar{\beta}(B) = \beta(B^{-1})$. This means that if $\lambda*\bar{\lambda} = Q$, then $Q = Q^2$, the support S_Q of Q is a compact group and $\mu*\bar{\mu} * Q = Q$, therefore, $F.F^{-1}.S_Q \subset S_Q$ which implies that F and

$$(X \times G \times Y)(X \times G_1 \times Y) = X \times G \times Y,$$

we have:

(41) $X \times G \times Y \subset \liminf_{n\to\infty}[F^n.(X \times G_1 \times Y)]$.

Since $F^n.(X \times G_1 \times Y) \subset X \times G \times Y$, in the definition of lim inf in (41), we can consider open sets in the relative topology of $X \times G \times Y$; in other words, for our purpose we will replace S by $X \times G \times Y$.

Let $(x,h,y) \in X \times G \times Y$. Then we can write:

$$(x,h,y) = (x,h_1,y)(x,h_2,y)$$

where $h_1,h_2 \in G$. Now given any open set U containing (x,h,y) we can find open sets U_1 containing (x,h,y) and U_2 containing (x,h_2,y) such that $U \supset U_1U_2$. Let us define:

$$V_2 = \{(x',z^{-1},y') : (x',z,y') \in U_2\}.$$

Since G is a topological group, V_2 is an open set. By (41), there is a positive integer N such that

$$U_1 \cap [F^N(X \times G_1 \times Y)] \neq \phi \text{ and}$$

$$V_2 \cap [F^N.(X \times G_1 \times Y)] \neq \phi$$

Let us define the set

$$A_N = \{(x',z^{-1},y'):(x',z,y') \in F^N.(X \times G_1 \times Y)\}$$

Then we have:

(42) $U_1U_2 \cap [F^N.(X \times G_1 \times Y).A_N \neq \phi$.

By (39),

(43) $F^N.(X \times G_1 \times Y) \subset X \times g^NG_1 \times Y$ ++

++the inclusion in (43) follows because we can write $X \times gG_1 \times Y$ as $(X \times G_1 \times Y).(X \times gG_1 \times Y)$ and then $F^2.(X \times G_1 \times Y) \subset F.(X \times gG_1 \times Y) = [F.(X \times G_1 \times Y)].(XxgG_1xY) \subset (XxgG_1xY).(XxgG_1xY) = X \times g^2G_1 \times Y$; the inclusion now follows by induction.

and therefore,

(44) $A_N \subset X \times g^{-N}G_1 \times Y.$

By (42), (43), and (44), we have:

$$U \cap (X \times g^N G_1 \times Y)(X \times g^{-N}G_1 \times Y) \neq \phi$$

or $\quad U \cap (X \times G_1 \times Y) \neq \phi$

Since U is an arbitrary open set containing $(x,h,y) \in X \times G \times Y$, this proves that $X \times G \times Y = X \times G_1 \times Y$ and so $G_1 = G$. Therefore (ii) => (iii).

(iii) => (i). This implication follows from the arguments given in the proof of Rosenblatt's Theorem 4.13.

Q.E.D.

Our next theorem gives necessary and sufficient conditions for the weak convergence of μ^n on a completely simple semigroup with its group factor compact. [Note that in this case weak convergence and weak*-convergence are equivalent].

4.18 Theorem. Let S be a locally compact completely simple semigroup with product representation $X \times G \times Y$, where G is compact. Let $\mu \in P(S)$, F be the support of μ and $S = \overline{\bigcup_{n=1}^{\infty} F^n}$. Then the following are equivalent:

(i) the sequence μ^n converge weakly

(ii) $\liminf_{n\to\infty} F^n$ is non-empty

(iii) there is no proper closed normal subgroup G_1 of G such that $YX \subset G_1$ and $F \subset X \times gG_1 \times Y$ for some g in $G - G_1$.

Proof. The proof is almost identical to that of Theorem 4.17 and is omitted.

Our next theorem gives a useful sufficient condition for

the convergence of μ^n on a compact semigroup or on a completely simple semigroup with compact group factor.

4.19. Theorem. Let S be a compact semigroup or a completely simple semigroup with compact group factor. Let $\mu \in P(S)$ and F be the support of μ. Then if k is the smallest positive integer such that $F^n \cap F^{n+k} \neq \phi$ for some positive integer n, then for each positive integer m with $0 \leq m \leq k-1$, the sequence μ^{m+nk} converge weakly on $n \to \infty$.

Proof. Let S be compact. [The proof in the other case will follow exactly similarly]. Let K_μ be the kernel of $\overline{\{\mu^n : n \geq 1\}}$ (weak-closure). Then K_μ is a group and K_μ consists only of all the cluster points of $\{\mu^n : n \geq 1\}$. Let η be the identity of K_μ. As in the proof of 4.13, it follows that S_η = the support of $\eta = X \times G_1 \times Y$, where $X \times G \times Y$ is the completely simple kernel of S, G_1 is a closed normal subgroup of G. Also, if λ is a cluster point of $\{\mu^n : n \geq 1\}$ other than η, then $S_\lambda =$ $X \times gG_1 \times Y$ where $g \in G - G_1$. It is also easy to show from the group property of K_μ that for any two distinct cluster points λ_1, λ_2, $S_{\lambda_1} \cap S_{\lambda_2} = \phi$. Since $\mu * \eta \in K_\mu$, it follows from the preceding discussion that

(45) $F.(X \times G_1 \times Y) = X \times gG_1 \times Y$ for some $g \in G$.

Then it follows from (45) that

(46) $F^n.(X \times G_1 \times Y) = X \times g^n G_1 \times Y$

for every positive integer n. Therefore if $F^n \cap F^{n+k} \neq \phi$, then $g^n G_1 \cap g^{n+k} G_1 \neq \phi$ which means $g^k G_1 = G_1$. This means that if k is the smallest positive integer with $F^n \cap F^{n+k} \neq \phi$ for some positive integer n, then k is the smallest positive integer for which

(47) $\mu^k * \eta = \eta$.

Hence for $0 \leq m \leq k-1$,

$$\mu^{m+nk} * \eta = \mu^m * \eta \text{ for all } n.$$

This means that if $\mu^{m+n_jk} \to \lambda_m \in P(S)$ weakly for some subsequence n_j of positive integers, then $\lambda_m = \lambda_m * \eta = \mu^m * \eta$. The theorem now follows.

Q.E.D.

In 4.13, we have found necessary and sufficient conditions for the convergence of μ^n on a compact semigroup. But the same problem for the sequence $\mu_1 * \mu_2 * \cdots * \mu_n$ where $\mu_n \in P(S)$, seems to be difficult and open, even when S is a compact group. However, the following theorem of Kloss [34] gives a sufficient condition for the convergence of the above sequence.

4.20. <u>Theorem</u>. Let S be a compact connected second countable group and (μ_n) be a sequence in $P(S)$. If there exists a pair of constants ε and δ, not depending on n, such that

(48)... $\mu_n(E) < 1 - \delta$ holds for any Borel set E for which $m(E) < \varepsilon$ (m = the normed Haar measure of S), then

$$\lim_{n\to\infty} \mu_1 * \mu_2 * \cdots * \mu_n = m.$$

<u>Proof</u>. It is easy to observe that

(i) if μ and υ satisfy (48), then $\mu * \upsilon$ satisfy (48); and

(ii) if the sequence υ_n (in $P(S)$) satisfy (48) and converge weakly to υ, then υ also satisfies (48).

Now consider the idempotent η in $\overline{E(\mu)} = \{\mu^n \mid n \geq 1\}^-$, where μ is a probability measure satisfying (48). Then η is the normed Haar measure on some compact subgroup H of S. If H is a proper subgroup of S, then $m(H) = 0$; since, $m(H) > 0$ implies $H.H^{-1} = H$

contains an open set and this implies that H is an open subgroup, and by the connectedness of S, H = S. Since by (i) above, each μ^n satisfies (48), it is clear that m(H) > 0 and therefore, $\eta = m$.

Now we write: $\upsilon_n = \mu_1 * \cdots * \mu_n$. By Theorem 4.16 (this theorem is also proven in [33] for compact groups), there exist $a_n \epsilon S$ such that $\upsilon_n^1 = \upsilon_n * \delta_{a_n}$ converges to some $\upsilon' \epsilon P(S)$. If we write: $\mu_1^1 = \mu_1 * \delta_{a_1}$, $\mu_n^1 = \delta_{a_{n-1}}-1 * \mu_n * \delta_{a_n}$ for n > 1, then $\upsilon_n^1 = \mu_1^1 * \ldots * \mu_n^1$.

Notice that if μ is a limit point for

$$\mu_{n+1}^1 * \ldots * \mu_{n+m}^1 \quad (n \to \infty),$$

then μ satisfies (48) and

$$\upsilon^1 = \upsilon^1 * \mu = \upsilon^1 * \mu^n \text{ for all n.}$$

This means that $\upsilon^1 = \upsilon^1 * m = m$. Hence $\upsilon_n * \delta_{a_n} \to m$ as $n \to \infty$. It is clear that $\upsilon_n \to m$ as $n \to \infty$.

Q.E.D.

Now we will study the convergence of the sequence $\upsilon_k^n = \mu_k * \ldots * \mu_n$ (as $n \to \infty$, in the weak topology) when S is a countable discrete group and the μ_i's are in P(S). This problem was earlier studied by Maximov for finite groups. He used a generalization of the classical concept of variance to find his results. Our approach to the problem is quite different and more elementary. We will show that Maximov's results continue to hold in infinite (discrete) groups, though his methods do not seem to carry over in the infinite case.

Even in the case of finite groups our results seem to be more complete than those of Maximov [44].

NOTE. In Propositions 4.21A and 4.21B as well as in Theorem 4.22 and its corollaries and the lemmas that are needed to prove this theorem, S is always a countable discrete group. In what follows, by Csiszar's theorem we will always mean either Theorem 4.29 or Proposition 4.21. The proofs of these two results of Csiszar are valid in any locally compact second countable group and do not depend upon any of our results in countable discrete groups. In fact, we'll use Csiszar's results often in proving Theorem 4.22.

First, we note that if $\nu_k^n \to \pi_k * w_G$ weakly as $n \to \infty$ for some k, then ν_p^n converges weakly for all $p \leq k$, but need not converge for any $p > k$. However, we have the following propositions.

4.21 Proposition (Csiszar [13]). Let T be a locally compact second countable group and $(\mu_i) \in P(T)$ such that every weak*-cluster point of ν_k^n is in P(T). Then there is a subsequence (n_i) of positive integers such that for all non-negative integers k, we have:

$$\nu_k^{n_i} \to \pi_k \in P(T) \text{ as } i \to \infty,$$

$$\pi_{n_i} \to \pi_\infty = \pi_\infty^2 \in P(T) \text{ as } i \to \infty \text{ and}$$

$$\pi_k * \pi_\infty = \pi_k.$$

4.21A Proposition. Suppose for some non-negative integer k, the sequence ν_k^n converges weakly to π_k and π_k is not of the form $\pi_k * w_G$ where $G \neq \{e\}$. Then for every non-negative integer k, the sequence ν_k^n converges weakly.

Proof. It is clear that ν_p^n converges weakly for all $p \leq k$. First, we notice that

$$(1') \qquad \liminf_{n\to\infty} \inf_{m>n} \nu_n^m(e) = 1.$$

This is because if (1′) is not true, then we can find sequences n_i, m_i

with $n_i < m_i$ of positive integers such that

$$\nu_{n_i}^{m_i} \to \lambda \quad \text{as } i \to \infty \text{ vaguely}$$

and $\lambda(e) < 1$. But since

$$\nu_k^{n_i} * \nu_{n_i}^{m_i} = \nu_k^{m_i} ,$$

we have:

$$\pi_k * \lambda = \pi_k \text{ and } \lambda \in P(S).$$

Then it follows easily that

$$\pi_k = \pi_k * w_G$$

where G is a finite group and G is generated by the support of λ. By the hypothesis of our proposition, $G = \{e\}$ and thus $\lambda = \delta_e$. Thus (1′) is valid. Now let $p > k$, and π_p', π_p'' be any two weak*-cluster points of ν_p^n. Then we can choose subsequences n_i and n_i' of positive integers such that $n_i + 1 < n'_i$ and

$$\nu_p^{n_i} \to \pi_p' \quad \text{and} \quad \nu_p^{n'_i} \to \pi_p''.$$

Since $\nu_p^{n'_i} = \nu_p^{n_i} * \nu_{n_i}^{n'_i}$, and by (1′) $\nu_{n_i}^{n'_i} \to \delta_e$, we have: $\pi_p' = \pi_p * \delta_e = \pi_p$. The proposition now follows.

A result that will be used in later results is the following.

4.21B <u>Proposition.</u> Suppose for every non-negative integer k, $\nu_k^n \to \pi_k$ weakly as $n \to \infty$. Then there is a finite subgroup G and a positive integer k_o such that for all k, $\pi_k = \pi_k * w_G$ and for all $k > k_o$, $\pi_k = \pi_k * w_H \Rightarrow H \subset G$.

<u>Proof.</u> Suppose $\nu_k^n \to \pi_k$ weakly for all k as $n \to \infty$. Then by Csiszar's Theorem there exists a subsequence n_i of the positive integers such that

$$\pi_{n_i} \to \pi_\infty = \pi_\infty^2 \in P(S) \text{ and } \pi_k * \pi_\infty = \pi_k,$$

for all non-negative integers k. Let G = the support of π_∞, which is a finite group. Let k_o be a positive integer such that $\pi_{k_o}(G) > \frac{1}{2}$. Let $k > k_o$. Then if $\pi_k = \pi_k * w_H$, then $\pi_{k_o} = \pi_{k_o} * w_H$. Hence $\pi_{k_o} = \pi_{k_o} * \delta_x$ for every $x \in H$. If $x^{-1} \in H - G$, then $\pi_{k_o}(G) = \pi_{k_o}(Gx^{-1})$, but $G \cap Gx^{-1} = \emptyset$ and therefore,

$$\pi_{k_o}(S) \geq \pi_{k_o}(G) + \pi_{k_o}(Gx^{-1}) > 1.$$

Hence for $k > k_o$, $\pi_k = \pi_k * w_H \Rightarrow H \subset G$. Q.E.D.

Our main theorem in this section is the following.

4.22 <u>Theorem.</u> Let G be a finite subgroup of S. Then the following statements are equivalent:

(a) $\sum_{n=1}^{\infty} \mu_n(S-G) < \infty$; and for any proper subgroup G' of G, there does not exist a sequence g_n, n = 0,1,2,... in S such that $\sum_{n=1}^{\infty} \mu_n(S - g_{n-1}G' g_n^{-1}) < \infty$.

(b) For all non-negative integers k, $\nu_k^n \to \pi_k$ weakly as

$n \to \infty$ and $\pi_k = \pi_k * w_G$; also there exists a positive integer k_o such that for all $k > k_o$, $\pi_k = \pi_k * w_H$ implies that $H \subset G$.

(c) For all non-negative integers k, $\nu_k^n \to \pi_k$ weakly as $n \to \infty$, $\pi_k = \pi_k * w_G$ and $\pi_n \to w_G$ as $n \to \infty$.

Before we prove this theorem, we present a number of interesting corollaries that follow from this theorem.

4.22A Corollary. For all non-negative integers k, the sequence $\nu_k^n \to \pi_k$ weakly as $n \to \infty$ if and only if there exists a finite subgroup G such that $\sum_{n=1}^{\infty} \mu_n(S-G) < \infty$ and for any proper subgroup G' of G and any selection of elements g_n in S, $n = 0,1,2,\ldots$, the series $\sum_{n=1}^{\infty} \mu_n(S - g_{n-1}G' g_n^{-1}) = \infty$.

Proof. The 'if' part follows from Theorem 4.22. For the 'only if' part, let $\nu_k^n \to \pi_k \in P(S)$ for all non-negative integers k. Then by Prop. 4.21B, the statement (b) of Theorem 4.22 follows and therefore, the corollary follows from Theorem 4.22.

4.22B Corollary. Suppose that all finite subgroups of S are normal subgroups. Then $\nu_k^n \to \pi_k$ weakly as $n \to \infty$ for all non-negative integers k if and only if there exists a finite subgroup G such that $\sum_{n=1}^{\infty} \mu_n(S-G) < \infty$ and for all proper subgroups G' of G, the series $\sum_{n=1}^{\infty} \inf\{\mu_n(S-xG'): x \in S\}$ is divergent.

We omit the proof of 4.22B which easily follows from 4.22.

4.22C Corollary. Suppose $\mu_n(e) = 1-r_n$. Then the series $\sum_{n=1}^{\infty} r_n < \infty$ if and only if the sequence $\nu_k^n \to \pi_k$ weakly as $n \to \infty$ for all non-negative integers k, and for some positive integer k_o, $k > k_o$ and H (a finite subgroup) $\neq \{e\}$ imply that $\pi_k \neq \pi_k * w_H$.

Proof. The corollary follows immediately from Theorem 4.22 by taking $G = \{e\}$ in this theorem.

The most interesting corollary is perhaps the following.

4.22D Corollary. Suppose S has no non-trivial proper finite subgroup. Let $\mu_n(e) = 1-r_n$. Then the following statements are true.

(a) If $\sum_{n=1}^{\infty} r_n < \infty$, then the sequence $\nu_k^n \to \pi_k$ weakly as $n \to \infty$ for all non-negative integers k.

(b) Suppose S is infinite and $\sum_{n=1}^{\infty} r_n = \infty$. Then ν_k^n does not converge weakly as $n \to \infty$ for any non-negative integer k.

(c) Suppose S is finite and $\sum_{n=1}^{\infty} r_n = \infty$. Then

(i) if $\sum_{n=1}^{\infty} [1 - \sup\{\mu_n(x): x \in S\}] < \infty$,

then for some positive integer k, the sequence ν_k^n does not converge weakly; and

(ii) if $\sum_{n=1}^{\infty} [1 - \sup\{\mu_n(x): x \in S\}] = \infty$, then for all non-negative integers k, the sequence $\nu_k^n \to \omega_S$ weakly as $n \to \infty$.

Proof. The statement (a) follows from 4.22C.

To prove (b), we use 5.4 which implies immediately, when $\sum_{n=1}^{\infty} r_n = \infty$, that for some positive integer k, the sequence ν_k^n does not converge weakly as $n \to \infty$. Now by Proposition 4.21A, it follows that the sequence ν_k^n cannot converge weakly for any positive integer k.

We now prove part (c). Assume that S is finite and $\sum_{n=1}^{\infty} r_n = \infty$. It follows that if G is a subgroup of S with $\sum_{n=1}^{\infty} \mu_n(S-G) < \infty$, then $G = S$. We also notice that

$$\sum_{n=1}^{\infty} \inf \mu_n(S-x\{e\}:x\epsilon S) = \sum_{n=1}^{\infty} [1-\sup\{\mu_n(x):x\epsilon S\}]$$

and therefore, if one of these series is divergent, then by Corollary 4.22B, it follows that for all non-negative integers k, $\nu_k^n \to w_S$ weakly as $n \to \infty$. If one of the above series is divergent, then again by Corollary 4.22B, ν_k^n does not converge weakly as $n \to \infty$ for some positive integer k. Q.E.D.

Now we will prove a sequence of lemmas which will be needed in the proof of Theorem 4.22

L1. Lemma. Suppose $\sum_{n=1}^{\infty} \mu_n(S-G) < \infty$ for some finite subgroup G of S. Then every weak*-cluster point of the sequence ν_k^n is a probability measure.

Proof. One way to prove this lemma is by using the classical Borel-Cantelli lemma. It follows easily that given $\epsilon > 0$, there is a positive integer k_o such that for any $k > k_o$ and all $n > k$, we have: $\nu_k^n(G) > 1 - \epsilon$. In other words, every weak*-cluster point of the double sequence ν_n^m (with $n \to \infty$ and $m > n$) must be a probability measure. The lemma now follows.

L2. Lemma. Let G be a subgroup of S and K any subset of S. Then for non-negative integers k, p and m with $k + 1 \leq p$ and $m \geq 1$, we have:

$$\nu_k^p(K.G) - \sum_{n=p+1}^{p+m} \mu_n(S-G) \leq \nu_k^{p+m}(K.G) \tag{2'}$$

$$\leq \nu_k^p(K.G) + \sum_{n=p+1}^{p+m} \mu_n(S-G).$$

Proof. By definition,

$$\nu_k^{p+1}(K.G) = \sum_{x\epsilon S} \nu_k^p(K.Gx^{-1})\mu_{p+1}(x)$$

$$\leq \nu_k^{p}(K.G) + \sum_{x \notin G} \nu_k^{p}(K.Gx^{-1})\mu_{p+1}(x)$$

$$\leq \nu_k^{p}(K.G) + \mu_{p+1}(S-G).$$

Repeating this process m times, we have:

$$\nu_k^{p+m}(K\cdot G) \leq \nu_k^{p}(K\cdot G) + \sum_{n=p+1}^{p+m} \mu_n(S-G).$$

Again,

$$\nu_k^{p+1}(K\cdot G) = \sum_{x\varepsilon S} \nu_k^{p}(KGx^{-1})\mu_{p+1}(x)$$

$$\geq \sum_{x\varepsilon G} \nu_k^{p}(KGx^{-1})\mu_{p+1}(x)$$

$$= \nu_k^{p}(KG)\,[1 - \mu_{p+1}(S-G)]$$

$$\geq \nu_k^{p}(KG) - \mu_{p+1}(S-G).$$

Repeating this process m times, we have:

$$\nu_k^{p}(KG) - \sum_{n=p+1}^{p+m} \mu_n(S-G) \leq \nu_k^{p+m}(KG).$$

The proof of the lemma is complete.

L3 . <u>Lemma</u>. Let G be a finite subgroup of S and $\sum_{n=1}^{\infty} \mu_n(S-G) < \infty$. Then for all non-negative integers k, the sequence $\nu_k^{n} * w_G \to \pi_k'$ weakly as $n \to \infty$ and $\pi_k' = \pi_k' * w_G$.

<u>Proof.</u> Let r be the number of elements in G. Then for any $x \varepsilon S$,

$$\nu_k^{n} * w_G(xG) = \nu_k^{n}(xG); \text{ also,}$$

$$\nu_k^n * w_G(x) = \nu_k^n * w_G(xg) \text{ for all } g \in G$$

so that we have:

(3′) $$\nu_k^n * w_G(x) = \frac{1}{r}\, \nu_k^n(xG).$$

Now by Lemma L2, the sequence $\nu_k^n(xG)$ (for fixed x in S) is a Cauchy sequence of real numbers since $\sum_{n=1}^{\infty} \mu_n(S-G) < \infty$. Hence by (3′), for every $x \in S$ the sequence $\nu_k^n * w_G(x)$ has a limit as $n \to \infty$. But this means that the sequence $\nu_k^n * w_G$ is weak*-convergent. By Lemma L1, it follows that this sequence is weakly convergent. Q.E.D.

L4 <u>Lemma</u>. If for every positive integer k, the sequence $\nu_k^n \to \pi_k$ weakly as $n \to \infty$ and if $\pi_{n_i} \to w_G$ weakly as $i \to \infty$ for some subsequence (n_i) of positive integers and some finite subgroup G, then the series $\sum_{n=1}^{\infty} \mu_n(S-G)$ is convergent.

<u>Proof</u>. Let $\varepsilon > 0$ and $\frac{1}{3} < \varepsilon$. Then there exists positive integers j and m with $m > j$ such that for all integers $p > m$, we have:

(4′) $$\nu_j^p(G) > 1 - \varepsilon > \frac{2}{3}.$$

Then we have:

$$\begin{aligned}\nu_j^{m+1}(G) &= \sum_{x \in S} \nu_j^m(Gx^{-1})\mu_{m+1}(x) \\ &\leq \nu_j^m(G)\mu_{m+1}(G) + (1 - \nu_j^m(G))(1 - \mu_{m+1}(G)) \\ &= \nu_j^m(G) - (1 - \mu_{m+1}(G))(2\nu_j^m(G) - 1) \\ &\leq \nu_j^m(G) - \tfrac{1}{3}(1 - \mu_{m+1}(G)), \text{ by (4′)}.\end{aligned}$$

Repeating this process, we have:

$$\nu_j^{m+p}(G) \leq \nu_j^{m}(G) - \frac{1}{3} \sum_{n=m+1}^{m+p} \mu_n(S-G).$$

It follows, by letting p tend to infinity, that

$$\sum_{n=m+1}^{\infty} \mu_n(S-G) \leq 3. \qquad \text{Q.E.D.}$$

L5 <u>Lemma</u>. Suppose for some non-negative integer k, the sequence ν_k^n has a non-zero weak*-cluster point λ_k. Then there exist elements $a_n \in S$ such that for all integers $p \geq k$, the sequence

$$\nu_p^n * \delta_{a_n} \to \pi_p \text{ weakly as } n \to \infty$$

such that $\pi_k = \lambda_k$.

<u>Proof</u>. By Csiszar's Theorem, there exist elements $b_n \in S$ such that for all positive integer $p \geq k$, we have:

$$\nu_p^n * \delta_{b_n} \to \pi_p \in P(S) \text{ as } n \to \infty.$$

Since $\nu_k^{n_i} \to \lambda_k \neq 0$ vaguely as $i \to \infty$, for some subsequence n_i of positive integers, the subsequence b_{n_i} must have a cluster point b. It follows that $\lambda_k * \delta_b = \pi_k$. We now define: $b_n b^{-1} = a_n$. Then

$$\nu_p^n * \delta_{a_n} \to \pi_p' = \pi_p * \delta_{b^{-1}}$$

weakly as $n \to \infty$ and $\pi_k' = \lambda_k$. Q.E.D.

Now we present the proof of Theorem 4.22

Proof of Theorem 4.22.

Step I. We first show that the statement (a) implies the statement (b). So we assume (a). By Lemma L1, there is a weak cluster point λ_k of $\nu_k{}^n$ in $P(S)$. By Csiszar's Theorem, there exists a subsequence (n_i) of positive integers such that for every non-negative integer k,

$$\nu_k{}^{n_i} \to \pi_k \in P(S) \quad \text{and} \quad \pi_{n_i} \to \pi_\infty = \pi_\infty{}^2$$

weakly as $i \to \infty$ and

$$\pi_k * \pi_\infty = \pi_k.$$

Let H be the support of π_∞. Then H is a finite group. We claim that $H = G$. Let $\varepsilon > 0$. Then there exists a positive integer N such that $\sum_{n=N}^{\infty} \mu_n(S-G) < \varepsilon$.

By Lemma L2 (taking $p = k + 1$ and $n = p + m$ in L2), we have: for all $k \geq N$ and $n \geq k + 2$,

$$\nu_k{}^n(G) \geq \mu_{k+1}(G) - \sum_{i=k+2}^{\infty} \mu_i(S-G)$$

$$\geq 1 - \varepsilon$$

and therefore, for all $k \geq N$

$$\pi_k(G) \geq 1 - \varepsilon.$$

This means that $\pi_\infty(G) \geq 1 - \varepsilon$. Since ε is arbitrary, this implies that $H \subset G$. To show that $G \subset H$, we note that there exists a positive integer i_0 such that for $i \geq i_0$, we have:

$$\pi_{n_i}(H) > \frac{1}{2}.$$

Now if $x \notin H$ and $\pi_{n_i} = \pi_{n_i} * \delta_x$, then $\pi_{n_i}(H) = \pi_{n_i}(Hx^{-1})$ implying $\pi_{n_i}(S) > 1$, a contradiction. Therefore, for all $i \geq i_o$,

$$\pi_{n_i} = \pi_{n_i} * w_{G'} \text{ for some finite subgroup } G' \Rightarrow G' \subset H.$$

Then if $n_i \leq k \leq n_{i+1}$ (for $i \geq i_o$), then

$$\pi_k = \pi_k * w_{G'} \text{ for some finite subgroup } G'$$

(5')

$$\Rightarrow \pi_{n_i} = \pi_{n_i} * w_{G'} \Rightarrow G' \subset H.$$

By Lemma L5, for some fixed k_o ($\geq n_{i_o}$) there exist elements $a_n \in S$ such that for all non-negative integers k

$$\nu_k^n * \delta_{a_n} \to \pi_k' \text{ weakly as } n \to \infty,$$

and $\pi_{k_o} = \pi_{k_o}'$.

We define:

$$\mu'_{k_o+1} = \mu_{k_o+1} * \delta_{a_{k_o+1}} \text{ and}$$

$$\mu_n' = \delta_{a_{n-1}^{-1}} * \mu_n * \delta_{a_n} \text{ for } n > k_o + 1.$$

Then we have: for $k = 0,1,2,\ldots,$

$$\nu_k'^n = \mu'_{k+1} * \ldots * \mu'_n = \nu_k^n * \delta_{a_n} \to \pi'_k$$

weakly as $n \to \infty$. Again by Csiszar's Theorem, there exists $\pi_\infty' = \pi_\infty'^2 \in P(S)$ such that

$$\pi_k' = \pi_k' * \pi_\infty', \ k = 0,1,2,\ldots$$

Now $\pi_\infty' = w_{G'}$ for some finite subgroup G'. Since $\pi_{k_0} = \pi_{k_0}'$, $\pi_{k_0} * w_{G'} = \pi_{k_0}$, and so by (5'), $G' \subset H$. But we have already proven that $H \subset G$ and so $G' \subset G$. By Lemma L4, we have:

$$\sum_{n=1}^{\infty} \mu_n'(S-G') < \infty, \text{ implying}$$

$$\sum_{n=1}^{\infty} \mu_n(S - a_{n-1}G'a_n^{-1}) < \infty.$$

By the hypothesis in (a), this is a contradiction unless G' = G. Hence G = H. Thus we have proven the following: for any non-negative integer k,

(i) By Lemma L3, $\nu_k^{\,n} * w_G$ converges weakly to some $\lambda \in P(S)$ as $n \to \infty$; and

(ii) If π_k is a weak cluster point of $\nu_k^{\,n}$, then $\pi_k = \pi_k * w_G$.

It follows that if π_k and π_k'' are weak cluster points of $\nu_k^{\,n}$, then

$$\pi_k'' = \pi_k'' * w_G = \lambda = \pi_k * w_G = \pi_k.$$

Hence the sequence $\nu_k^{\,n}$ converges weakly as $n \to \infty$. The rest of the statement (b) now follows from (5).

<u>Step II</u>. Now we show that (b) implies (c). Let λ be a vague cluster point of the sequence (π_n). Then it is clear that λ is a vague cluster point of the double sequence $\nu_n^{\,m}$ (with $n \to \infty$ and $m > n$). Since $\nu_k^{\,n} * \nu_n^{\,m} = \nu_k^{\,m}$, it follows that $\pi_k * \lambda = \pi_k$, implying that λ is a probability measure. Since λ is a cluster point of the π_n and for each k, $\pi_k = \pi_k * w_G$, it follows that

$$\lambda * w_G = \lambda = \lambda * \lambda.$$

Therefore, $\lambda = w_H$ for some finite subgroup H. Now for $k > k_o$, $\pi_k * w_H = \pi_k$ and therefore by the assumption of (b), $H \subset G$. Therefore,

$$\lambda = \lambda * w_G = w_G.$$

Thus (c) follows.

Step III. To complete the proof of the theorem we now show that (c) implies (a). So we assume that for $k = 0,1,2,\ldots$

$$\nu_k^n \to \pi_k \text{ weakly as } n \to \infty, \text{ and}$$

$$\pi_n \to w_G \text{ as } n \to \infty,$$

where G is a finite group. By Lemma L4,

$$\sum_{n=1}^{\infty} \mu_n(S-G) < \infty.$$

We will prove (a) by contradiction. Suppose there exists a proper subgroup G' such that there are elements $a_n \in S$, $n = 0,1,2,\ldots$, satisfying

$$\sum_{n=1}^{\infty} \mu_n(S - a_{n-1}G'a_n^{-1}) < \infty.$$

Since there are only finitely many subgroups of G, we can assume with no loss of generality that G' does not have a proper subgroup with the above property. Now μ_n' is defined by:

$$\mu_n' = \delta_{a_{n-1}^{-1}} * \mu_n * \delta_{a_n}.$$

Then

$$\sum_{n=1}^{\infty} \mu_n'(S-G') = \sum_{n=1}^{\infty} \mu_n(S - a_{n-1}G'a_n^{-1}) < \infty,$$

and for all proper subgroups G" of G' and elements $g_n \in S$, $n = 0,1,\ldots$

$$\sum_{n=1}^{\infty} \mu_n'(S - g_{n-1}G''g_n^{-1})$$

$$= \sum_{n=1}^{\infty} \mu_n(S - a_{n-1}g_{n-1}G''g_n^{-1}a_n^{-1})$$

$$= \infty$$

by the way G' was chosen. Since "(a) implies (b)" has already been proven, it follows that for all $k = 0,1,2,\ldots,$

$$\nu_k'^{\,n} = \mu'_{k+1} *\ldots* \mu'_n \to \pi'_k \text{ weakly as } n \to \infty,$$

$$\pi'_k = \pi_k' * w_{G'}, \quad \text{and}$$

there exist a positive integer k_o such that $k > k_o$ and

$$\pi_k' = \pi_k' * w_H \text{ implies that } H \subset G'.$$

Since $\nu'^{\,n}_k = \delta_{a_{k-1}^{-1}} * \nu_k^n * \delta_{a_n}$ and ν_k^n as well as $\nu'^{\,n}_k$ converge weakly, it is clear that the sequence a_n has a cluster point $b \in S$. Then

$$\pi'_k = \delta_{a_{k-1}^{-1}} * \pi_k * \delta_b.$$

Now for all non-negative integers k,

$$\pi'_k * \delta_{b^{-1}} * w_G * \delta_b = \delta_{a_{k-1}^{-1}} * \pi_k * w_G * \delta_b$$

$$= \delta_{a_{k-1}^{-1}} * \pi_k * \delta_b \ (\text{since } \pi_k = \pi_k * w_G) \quad = \pi_k'.$$

But this means that

$$b^{-1}\ G\ b \subset G',$$

which is a contradiction since G' is a proper subgroup of the finite group G. Q.E.D.

Our next theorem gives a necessary and sufficient condition for the convergence of $\sup\{\nu_1^n(Kx) : x \in S\}$ to 0 as n tends to infinity for every finite set K in a countable discrete group S.

4.22E <u>Theorem</u>. Let S be a countable discrete group and (μ_i) be a sequence in P(S). Then there exist elements a_n in S such that the sequence $\nu_k^n * \delta_{a_n}$ converges weakly as n tends to infinity for all nonnegative integers k if and only if there exists a finite subgroup G of S such that $\Sigma\, \mu_n(S-g_{n-1}Gg_n^{-1}) < \infty$ for <u>some</u> selection of elements g_n in S with $g_o=e$.

The proof of this theorem follows easily from the above results by considering the sequence $\mu_n' = \delta_{g_{n-1}^{-1}} * \mu_n * \delta_{g_n}$ for the 'if' part and the sequence $\mu_n' = \delta_{a_{n-1}^{-1}} * \mu_n * \delta_{a_n}$ for the 'only if' part. We omit the proof.

The above results can also be studied and obtained in a more general context while studying the almost sure convergence of products of independent random variables with values in a discrete completely simple semigroup. Earlier, relationship between almost sure convergence and convergence in distribution for such random variables were studied by Loynes [40], Csiszar [13] and Galmarino [17] in the case of groups and by Ito and Nishio [29] in the case of Banach spaces. In what follows, we will present results extending most of the results of the above authors. Though some of our results (the next three theorems) are generalizable to the locally compact case, for the sake of simplicity we'll restrict ourselves to the discrete case.

For probability measures $x_1, x_2, \ldots$, we will write occasionally: $x_n^m = x_n * x_{n+1} * \ldots * x_m$.

Our first theorem concerning almost sure convergence of products of independent semigroup-valued random variables is the following.

4.23 Theorem. Let $X_1(w), X_2(w), \ldots$ be a sequence of independent random variables taking values in a discrete completely simple semigroup $S = E \times G \times F$. Then the following conditions are equivalent:

(i) the sequence $Z_n(w) = X_1(w)X_2(w)\ldots X_n(w)$ converges almost surely to $Z(w)$ as $n \to \infty$;

(ii) there exists $f \in F$ such that $\sum_{n=1}^{\infty} r_n < \infty$, where $r_n = 1 - x_n(I_f)$ and I_f is the set of idempotents in $E \times G \times \{f\}$.

Proof. First, let us assume (i) and establish (ii). Suppose $Z_n(w)$ converges a.s. to $Z(w)$ as $n \to \infty$. Then for some $f \in F$,

(52) $P(Z(w) \in E \times G \times \{f\}) = \delta > 0.$

Notice that

(53) $\{w : Z(w) \in E \times G \times \{f\}\} =$

$$\bigcup_{n=1}^{\infty} \bigcap_{k=n}^{\infty} \{w : Z_k(w) \in E \times G \times \{f\}\}.$$

Given $\varepsilon > 0$, we can find positive integers m and n such that

(54) $$\delta - \varepsilon \leq P(\bigcap_{k=n}^{\infty} (Z_k(w) \in E \times G \times' \{f\}))$$

$$\leq P(\bigcap_{k=n}^{m} (Z_k(w) \in E \times G \times' \{f\}))$$

$$\leq \delta + \varepsilon.$$

Now we note that

(55) $Z_k(w) \in E \times G \times \{f\} \iff X_k(w) \in E \times G \times \{f\}.$

From (54) and (55),

$$\delta - \varepsilon \leq P(\bigcap_{k=n}^{m} (Z_k(w) \in E \times G \times \{f\})$$

$$\bigcap (X_j(w) \in E \times G \times \{f\} \text{ for all } j > m))$$

$$= P(\bigcap_{k=n}^{m} (Z_k(w) \in E \times G \times' \{f\}))$$

$$P(\bigcap_{j=m+1}^{\infty} (X_j(w) \in E \times G \times' \{f\}))$$

$$\leq (\delta+\varepsilon)\cdot\delta.$$

Since $\varepsilon > 0$ is chosen arbitrarily, this means that $\delta = 1$.

Now we observe that

$$Z_n(w) = Z_{n+1}(w) \in E \times G \times' \{f\} \text{ only if}$$

$X_{n+1}(w) \in I_f$. Therefore, we have from the fact that

$$P(\bigcup_{k=1}^{\infty} \bigcap_{n=k}^{\infty} \{Z_n(w) = Z(w) \in E \times G \times' \{f\}\}) = 1,$$

the following assertion:

$$(56) \quad P(\bigcup_{k=1}^{\infty} \bigcap_{n=k}^{\infty} \{X_n(w) \in I_f\}) = 1.$$

From (56), we have:

$$P(X_n(w) \in I_f^{\,c} \text{ infinitely often}) = 0.$$

By Borel-Cantelli lemma, it follows that

$$(57) \quad \sum_{n=1}^{\infty} P(X_n(w) \in I_f^{\,c}) < \infty,$$

establishing (ii).

The converse is obvious, since the condition (ii) implies (56) by the converse part of the Borel-Cantelli lemma and the validity of (56), in turn, establishes that the sequence $Z_n(w)$ converges almost surely. Q.E.D.

Our next theorem shows how almost sure convergence and convergence in distribution are related for products of independent random variables.

First, we present a simple example which will be useful in the context of our next result. This example will also show, among other things, the essential difference between the group case and the semigroup case in the context of the <u>equivalence theorem</u>. See Loeve [39].

4.24 <u>Example</u>. Consider the finite completely simple semigroup $S = G \times F$ where $G = \{u\}$ and $F = \{f_1, f_2\}$. This is a right-zero semigroup and the multiplication is defined as:

$$(u,f_1)(u,f_2) = (u,f_2) = (u,f_2)(u,f_2)$$

and $(u,f_2)(u,f_1) = (u,f_1) = (u,f_1)(u,f_1)$

Here I_{f_1} = the set of idempotents in $G \times \{f_1\} = \{(u,f_1)\}$. Let us define:

$$x_n(\{(u,f_1)\}) = \frac{n-1}{n}, \quad x_n(\{(u,f_2)\}) = \frac{1}{n}$$

Then since

$$I_{f_1} s^{-1} = \{y \in S \mid ys \in I_{f_1}\}$$

is empty when $s \notin I_{f_1}$ and is S when $s \in I_{f_1}$, we have: for $m > n$,

$$x_n^m(I_{f_1}) = x_n * x_{n+1} * \ldots * x_m(I_{f_1})$$

$$= \sum_{s \in I_{f_1}} x_n^{m-1}(I_{f_1} s^{-1}) x_m(s)$$

$$= x_m(I_{f_1}) = \frac{m-1}{m} .$$

It is clear that

(i) $\liminf_{n \to \infty \; m > n} x_n^m(I_{f_1}) = 1$, and

(ii) for each positive integer k, the sequence x_k^n converges to the unit mass at (u,f_1).

Clearly, it follows from Theorem 3.1 that $Z_n(w)$ does not

converge almost surely in this case.

The situation in the case of group-valued random variables, is different. It will follow from the next theorem that in the case of a group, almost sure convergence of $Z_n(w)$ is equivalent to the condition: $\lim_{n\to\infty} \inf_{m>n} x_n^m(u) = 1$ where u is the identity of the group.

4.25 Theorem. Let $X_1(w), X_2(w),\ldots$ be a sequence of independent random variables taking values in a discrete completely simple semigroup $S = E \times G \times F$. Then the following conditions are equivalent:

(i) the sequence $Z_n(w)$ converges almost surely;

(ii) there exists $f \in F$ such that $\sum_{n=1}^{\infty} R_n < \infty$, $R_n = 1 - x_n(E \times G \times' \{f\})$ and $\lim_{n\to\infty} \inf_{m>n} x_n^m(I_f) = 1$, where I_f is the set of idempotents in $E \times G \times' \{f\}$; and

(iii) there exists $f \in F$ such that $\sum_{n=1}^{\infty} R_n < \infty$; moreover, for each positive integer k, the sequence x_k^n converges weakly to some probability measure $x^{(k)}$ with its support contained in $E \times G \times \{f\}$ and for some k, the projection of $x^{(k)}$ on the group $G_1 = \{e\} \times G \times \{f\}$ is not of the form $\beta * w_H$ where $\beta \in P(G_1)$ and H is a finite subgroup of G_1 containing more than one element. [Here w_H stands for the uniform or Haar probability measure on H.]

Proof. First, we remark that the assertion (i) => (ii) follows immediately from Theorem 3.1 because of the identity (56) in its proof.

To show that (ii) implies (i), we notice that

$$\lim_{n\to\infty} P\Big({\bigcap_{k=n}^{\infty}}' \{Z_k(w) \in E \times G \times' \{f\}\}\Big)$$

$$= \lim_{n\to\infty} P\left(\bigcap_{k=n}^{\infty} \{X_k(w) \in E \times G \times' \{f\}\}\right)$$

$$= \lim_{n\to\infty} \prod_{k=n}^{\infty} x_k(E \times G \times \{f\})$$

$$= 1, \text{ since } \sum_{n=1}^{\infty} R_n < \infty \text{ by (ii)}.$$

By (ii), given $\varepsilon > 0$, there exists a positive integer n_o such that for all $m > n \geq n_o$, we have

(58) $\quad x_n{}^m(I_f) \geq 1 - \varepsilon,$

$$\text{i.e. } P(X_n(w)X_{n+1}(w)\ldots X_m(w) \in I_f) \geq 1 - \varepsilon$$

and also

$$P(Z_n(w) \in E \times G \times \{f\} \text{ for all } n \geq n_o) \geq 1 - \varepsilon.$$

This means that for all $m > n \geq n_o$,

$$P(Z_n(w) = Z_m(w)) \geq 1 - 2\varepsilon.$$

Note that for two elements (e,g,f) and (e_1,g_1,f), if $(e,g,f)\cdot(e_1,g_1,f) \in I_f$, then $(e,g,f) \notin I_f$ implies that $(e_1,g_1,f) \notin I_f$. Using (58), we have for $m > n \geq n_o$,

$$P(Z_n(w) \neq Z_{n+i}(w) \text{ for some } i,\ 1 \leq i < m - n)$$

$$\leq P(Z_n(w) = Z_m(w),\ Z_k(w) \in E \times G \times \{f\}$$
$$\text{for all } k \geq n_o,\ Z_n(w) \neq Z_{n+i}(w) \text{ for some } i,$$
$$1 \leq i < m - n) + 2\varepsilon$$

$$\leq \sum_{i=1}^{m-n-1} P(X_{n+1}(w)\ldots X_{n+j}(w) \in I_f,\ 1 \leq j < i,$$
$$X_{n+1}(w)\ldots X_{n+i}(w) \notin I_f,\ X_{n+i+1}(w)\ldots X_m(w) \notin I_f) + 2\varepsilon$$

$$= \sum_{i=1}^{m-n-1} P(X_{n+1}(w)\ldots X_{n+j}(w) \in I_f, 1 \leq j < i, X_{n+1}(w)\ldots X_{n+i}(w) \notin I_f)\cdot$$
$$P(X_{n+i+1}(w)\ldots X_m(w) \notin I_f) + 2\varepsilon.$$

$$\leq \varepsilon + 2\varepsilon = 3\varepsilon, \text{ by (58)}.$$

Letting m tend to ∞, it follows that

$$P\{w : \text{there exists a positive integer } n \text{ such that } Z_n(w) = Z_m(w) \text{ for every } m > n\} = 1.$$

Hence $Z_n(w)$ converges almost surely. Thus (i) $\Leftarrow$ (ii).

Now we show that (ii) $\Leftarrow$ (iii). Since (ii) $\Leftarrow$ (i), it follows by (ii) that for every positive integer k, the sequence $X_k(w)\ldots X_n(w) = Z_k{}^n(w)$ converges almost surely and hence $x_k{}^n$, the distribution of $Z_k{}^n(w)$, converges weakly to $x^{(k)}$, the distribution of $\lim_{n\to\infty} Z_k{}^n(w)$ and the support of $x^{(k)}$ is contained in $E \times G \times' \{f\}$.

Now suppose that for each positive integer k, the projection μ_k of $x^{(k)}$ on the group $G_1 = \{e\} \times G \times' \{f\}$, $e \in E$, is $\beta_k * w_{H_k}$ where H_k is a finite subgroup of G_1 containing more than one element. Then by (ii), it follows that there exists a positive integer n_o such that for all $m > n_o + 1$,

$$\mu_{n_o}{}^m(u) > \frac{2}{3}, \ \mu_{n_o}{}^m = \text{the projection of } x_{n_o}{}^m \text{ on } G_1 \text{ and}$$

$$u = \text{the identity of } G_1.$$

This means that $\mu_{n_o}(u) > \frac{2}{3}$. But since $\mu_{n_o} = \beta_{n_o} * w_{H_{n_o}}$, for any element $h \in H_{n_o}$, $h \neq u$, $\mu_{n_o}(u) = \mu_{n_o}(h) > \frac{2}{3}$. This is a contradiction, since $\mu_{n_o}(G_1) = 1$. Thus (iii) follows.

Now we assume (iii) and prove (ii). By (iii), the projections $\mu_k{}^n$ of x_k^r on G_1 converge weakly to the projection μ_k of $x^{(k)}$ on G_1. This means that if υ is a cluster point of the double sequence $\mu_n{}^m$ ($m > n$ and $n \to \infty$), then $\mu_k * \upsilon = \mu_k$. [Note that in a group, the sequence $\mu_n{}^m$ can have only probability measures as its cluster points, since the sequence $\mu_k{}^n$ is weakly convergent.] This means that

$$\mu_k * \left(\left(\frac{1}{p}\right) \sum_{n=1}^{p} \upsilon^n\right) = \mu_k \text{ for every } p.$$

Since $\left(\frac{1}{p}\right) \sum_{n=1}^{p} \upsilon^n$ converges to the uniform measure of a subgroup H of G_1 (which can be verified easily)

$$\mu_k * w_H = \mu_k.$$

By (iii), H = {u} and consequently, υ = the unit mass at u. This means that

$$\lim_{n\to\infty} \inf_{m>n} \mu_n^m(u) = 1.$$

Now (ii) follows immediately. The proof of the theorem is complete.

Now we discuss the structure of the weak limit z of the sequence of probability measures $x_1^n = x_1 * x_2 * \ldots * x_n$ where $x_i \in P(S)$ and S is a discrete completely simple semigroup. It is clear that if $\beta \in P(S)$ and $x_1 = \beta$, $x_n = I = I * I \in P(S)$ for all $n > 1$, then $x_1^n \to \beta * I$. In the next theorem we'll show that in most cases, the limit measure z is of the form $\beta * I$, where $\beta \in P(S)$ and I is an idempotent probability measure on S.

4.26 Theorem. Let S = E x G x F be a discrete completely simple semigroup with E finite. Let $x_i \in P(S)$, $1 \le i < \infty$. Then if $z \in P(S)$ and $x_1^n \to z$ weakly as $n \to \infty$, then $z = \beta * I$, where $\beta \in P(S)$ and $I = I * I \in P(S)$.

Proof. Suppose $x_1^n \to z$ weakly as $n \to \infty$. Consider the double sequence x_n^m, $m > n$ and $n \to \infty$. We claim that this sequence is conditionally compact in the weak topology of P(S). Let $\varepsilon > 0$. Let K be a finite subset of S such that $z(K) > 1 - \varepsilon$. Then $\exists$ n_0 such that for all $n \ge n_0$, $x_1^n(K) > 1 - \varepsilon$. Let $m > n \ge n_0$. Then

$$1 - \varepsilon < x_1^{\,m}(K) = \sum_{s\varepsilon S} x_{n+1}^{\,m}(s^{-1}K)x_1^{\,n}(s)$$

$$\leq \sum_{s\varepsilon K} x_{n+1}^{\,m}(s^{-1}K)x_1^{\,n}(s) + \varepsilon$$

$$\leq x_{n+1}^{\,m}(K^{-1}K) + \varepsilon.$$

Now noting that $K^{-1}K$ is finite since E is finite, our claim is verified and therefore, the double sequence $x_n^{\,m}$ can have only probability measures as its cluster points. Let υ be a cluster point. Then since

$$x_1^{\,n} * x_{n+1}^{\,m} = x_1^{\,m},$$

we have: $z * \upsilon = z$. This means that $z * \upsilon_n = z$ where $\upsilon_n = \left(\frac{1}{n}\right)\sum_{k=1}^{n} \upsilon^k$. Note that given $\varepsilon > 0$, there is a finite set K such that $z(K) > 1 - \varepsilon$ and therefore,

$$1 - \varepsilon < z(K) = \sum_{s\varepsilon S} \upsilon_n(s^{-1}K)z(s)$$

$$\leq \sum_{s\varepsilon K} \upsilon_n(s^{-1}K)z(s) + \varepsilon$$

$$\leq \upsilon_n(K^{-1}K) + \varepsilon.$$

This means that the sequence υ_n is weakly conditionally compact and therefore, by the same proof as given for Prop. 4.3, we have

$$\upsilon_n \to I = I * I \;\varepsilon\; P(S), \text{ as n tends to infinity.}$$

Hence $z * I = z$ and the theorem is proved.

4.27 Corollary. In a finite completely simple semigroup, the limit z of the sequence $x_1^{\,n}$ (when it converges) is of the form $\beta * I$, where β and I are probability measures and $I = I * I$.

We conjecture that Theorem 4.26 holds even without the assumption that E is finite. We are unable to prove this at this time.

4.28 Remarks. We remark that some of the preceding theory carries over in the locally compact (non-discrete) situation. In the case of locally compact groups, it is possible to obtain complete results along the lines of our results in 4.23-4.26. However, in the case of completely simple non-discrete semigroups some problems (mostly measure-theoretic) come up. For example, if the sequence x_1^n is weakly convergent, then it is not true that the double sequence x_n^m can have only probability measures as weak cluster points. This can be shown by considering a left group S with an infinite set $(e_n)_{n=1}^{\infty}$ of idempotents. Then if x_n is the unit mass at e_n, it is clear that x_1^n = the unit mass at e_1 where as x_n^m = the unit mass at e_n converge vaguely to 0 as $n \to \infty$, if e_n's don't have a cluster point.

Now we state a useful theorem first proven by Csiszar [13], see also Tortrat [75] for locally compact second countable groups. This theorem was proven by Kloss [33] for compact groups. Since we will not use the method of Csiszar in what follows (note that we have already used Csiszar's method of tail idempotents in 4.14 - the same method is followed in [13]), the proof of Csiszar-Tortrat's Theorem will be omitted.

4.29 Theorem. Let S be a locally compact second countable topological group and (μ_n) be a sequence in P(S). Then either

$$\sup \{\mu_1 * \mu_2 * \ldots * \mu_n(Kx) : x \in S\}$$

goes to zero as $n \to \infty$ for every compact K or there exist elements $a_n \in S$ such that for each positive integer k, the sequence

$$\mu_k * \mu_{k+1} * \ldots * \mu_n * \delta_{a_n}$$

converges weakly to a probability measure as $n \to \infty$.

In the next few results (4.31-4.35) S _will always denote a locally compact non-compact topological group_. We know from 4.14 that if $\mu \in P(S)$ and S is the smallest closed group containing S, then $\mu^n \to o$ vaguely as $n \to \infty$. We will show now that actually a stronger form of this result is valid in many topological groups.

4.30 _Proposition_. Let $\mu \in P(S)$. Suppose there exist $a_n \in S$ such that the sequence $\mu^n * \delta_{a_n}$ converges vaguely as $n \to \infty$ to a probability measure $Q \in P(S)$. Then for some $z \in S$, $\mu * Q = Q * \delta_z$.

Proof. If $\mu^n * \delta_{a_n} \to Q$ as $n \to \infty$, then $\mu^{n+1} * \delta_{a_n} \to \mu * Q$ as $n \to \infty$. But $\mu^{n+1} * \delta_{a_n} = \mu^{n+1} * \delta_{a_{n+1}} * \delta_{a_{n+1}^{-1} a_n}$. This means that if $a_{n+1}^{-1} a_n \to \infty$, then $\mu^{n+1} * \delta_{a_n} \to o$ as $n \to \infty$ which is impossible. Therefore, $a_{n+1}^{-1} a_n \not\to \infty$ and consequently, this sequence has some limit point $z \in S$. It follows easily that $\mu * Q = Q * \delta_z$.

Q.E.D.

4.31 _Theorem_. Suppose $\mu \in P(S)$ and $S = \overline{\bigcup_{n=1}^{\infty} S_\mu^n}$. Suppose there exists an open set V with compact closure such that for every $x \in S$, $x^{-1}Vx = V$. Then for every compact K, $\sup\{\mu^n(Kx) : x \in S\} \to o$ as $n \to \infty$.

Proof. Suppose the conclusion of the theorem is false. Then by Csiszar's theorem 4.29, $\mu^n * \delta_{a_n} \to Q \in P(S)$ as $n \to \infty$. With no loss of generality, we can assume that $S_Q \cap V \neq \phi$. Now by Prop.4.30, $\mu * Q = Q * \delta_z$ for some $z \in S$. Then $\mu^n * Q = Q * \delta_z n$.

The function $x \to Q(x^{-1} K)$, $K = \overline{V}$, is upper semi-continuous and therefore, attains its maximum at some $x = x_o$ in S. Then

$$Q(\bar{x}_o^1 K) = \mu^n * Q(\bar{x}_o^1 K z^n)$$

$$= \int \mu^n (\bar{x}_o^1 K z^n y^{-1}) Q(dy)$$

$$= \int \mu^n (\bar{x}_o^1 \bar{z}^n \bar{y}^1 K) Q(dy)$$

$$= \int_{\bar{z}^n x_o} (\mu^n) (\bar{y}^1 K) Q(dy)$$

where for any measure λ, ${}_x\lambda(B) = \lambda(x^{-1} B)$

Let us write: $\upsilon = \sum_{n=1}^{\infty} \frac{1}{2}^n {}_{\bar{z}^n x_o}(\mu^n)$. Then υ is a probability measure and we have:

$$Q(x_o^{-1}K) = \sum_{n=1}^{\infty} \frac{1}{2}^n Q(x_o^{-1} K)$$

$$= \int \upsilon (y^{-1} K) Q(dy)$$

$$= \int \upsilon (Ky^{-1}) Q(dy)$$

$$= \upsilon * Q(K) = \int Q(y^{-1} K) \upsilon(dy)$$

This means that

$$\int [Q(x_o^{-1}K) - Q(y^{-1}K)] \upsilon(dy) = 0.$$

Since $Q(x_o^{-1}K) \geq Q(y^{-1}K)$ for all $y \in S$, it follows that $Q(x_o^{-1}K) = Q(y^{-1}K)$ for almost all $y(\upsilon)$ in S_υ. By the upper semicontinuity of the function $x \to Q(x^{-1}K)$, $Q(x_o^{-1}K) = Q(y^{-1}K)$ for <u>all</u> $y \in S_\upsilon$. This means that S_υ is compact. To see this, suppose there exist infinitely many $y_n \in S$ such that the sequence y_n doesn't have a limit point. Then for each n, $Q(y_n^{-1}K) = Q(x_o^{-1}K) > 0$; also since $y_n^{-1}K \cap y_1^{-1}K \neq \phi$ for all n implies that $y_n \in K(y_1^{-1}K)^{-1}$ (which is a compact set) for all n, there exists $n_1 > 1$ such that $y_n^{-1}K \cap y_1^{-1}K = \phi$. Again, since $y_n^{-1}K \cap (y_{n_1}^{-1}K \cup y_1^{-1}K) \neq \phi$ for all $n > n_1$ implies that the y_n's (for all $n > n_1$) lie in a compact set, there exists $n_2 > n_1$ such that $y_{n_2}^{-1}K \cap (y_{n_1}^{-1}K \cup y_{n_1}^{-1}K) = \phi$. In this way, we can show the existence of infinitely many pairwise disjoint sets $y_{n_i}^{-1}K$, each having the same positive Q-measure. This contradicts that Q is a probability measure, proving that S_υ is compact.

Now we claim that $H = \overline{\bigcup_{k=1}^{\infty} S_\mu^{-k} S_\mu^{k}}$ is a compact normal subgroup of S. To see this, we notice that $S_\upsilon = \overline{\bigcup_{k=1}^{\infty} z^{-k} x_o S^k_\mu}$, and $S_\upsilon^{-1} S_\upsilon \supset H$ so that H is compact. Since by hypothesis, $S = \overline{\bigcup_{k=1}^{\infty} S_\mu^{k}}$, it is clear that for any $x \in S$, $x^{-1} Hx \subset H$, this inclusion being easily valid for all $x \in S_\mu^{n}$ (n any positive integer). This means that for all $x \in S$, $xH = Hx$. It follows that H is a subsemigroup since for any two positive integers m and n,

$$
\begin{aligned}
&(S_\mu^{-m} S_\mu^{m})(S_\mu^{-n} S_\mu^{n}) \\
&\subset H (S_\mu^{-n} S_\mu^{n}) = (H S_\mu^{-n}) S_\mu^{n} \\
&= (S_\mu^{-n} H) S_\mu^{n} \\
&\subset H.
\end{aligned}
$$

Since $H = H^{-1}$, H is a compact normal subgroup of S as claimed.

Since $S_\mu^{-1} S_\mu \subset H$, it follows immediately that $S_\mu \subset H \cdot x$ for some x. This $x \notin H$, since otherwise $S_\mu \subset H$ = a compact subgroup, contradicting that the smallest closed subgroup containing S is non-compact. Since $S_\mu^{n} \subset Hx^n$, by the normality of H, $S = \overline{\bigcup_{n=1}^{\infty} S_\mu^{n}} = \overline{\bigcup_{n=1}^{\infty} Hx^n}$. Now we notice that the set $\bigcup_{n=1}^{\infty} Hx^n$ is closed. The reason is: Let y be not contained in, but a limit point of $\bigcup_{n=1}^{\infty} S_\mu^{n}$; then there are elements $u_{n_i} \in S_\mu^{n_i}$ such that $u_{n_i} \to y$ as $i \to \infty$. Notice that $S_\upsilon = \overline{\bigcup_{k=1}^{\infty} z^{-k} x_o S_\mu^{k}}$ is compact. Hence there is a subsequence n_{i_j} such that

$z^{-n_{i_j}} \cdot x_o u_{n_{i_j}} \to w \in S$. But this means that $z^{n_{i_j}} \to x_o \cdot yw$ as $j \to \infty$. Since by Proposition 4.30, $\mu^{n_{i_j}} * Q = Q * \delta_{z^{n_{i_j}}}$, this means that $\mu^{n_{i_j}} * Q \to Q * \delta_{x_o yw}$ as $j \to \infty$. This is a contradiction, since $\mu^n \to 0$ vaguely as $n \to \infty$.

It is now clear that $S = \bigcup_{n=1}^{\infty} Hx^n$. Let $a \in S_\mu \subset Hx$. Then $a^{-1} \in Hx^n$ for some positive integer n and therefore, $Hx^n = Hx^{-1}$ or $x^{n+1} \in H$. This means that $S = \bigcup_{k=1}^{n} Hx^k$, which is a contradiction since S is non-compact. The theorem now follows.

The following corollary now follows immediately.

4.32 <u>Corollary</u>. Let S be discrete, $\mu \in P(S)$ and $S = \bigcup_{k=1}^{\infty} S_\mu^k$. Then for every compact set K, $\sup\{\mu^n(Kx) : x \in S\} \to 0$ as $n \to \infty$.

It is <u>not</u> clear how we can prove Theorem 4.31, if, instead of assuming $S = \overline{\bigcup_{k=1}^{\infty} S_\mu^k}$ (i.e. the group S is generated by S_μ as a semigroup), we assumed $S = \overline{\bigcup_{k=1}^{\infty} (S_\mu \cup S_\mu^{-1})^k}$ (i.e. the group S is generated by S_μ as a group). The difficulty is in the showing (in the proof of Theorem 4.31) that $H = \overline{\bigcup_{k=1}^{\infty} S_\mu^k S_\mu^{-k}}$ is a semigroup. However, we can prove the following theorem when S is abelian or μ is symmetric.

4.33 <u>Theorem</u>. Suppose $\mu \in P(S)$ and $S = \overline{\bigcup_{k=1}^{\infty} (S_\mu \cup S_\mu^{-1})^k}$. Suppose also that <u>either</u> S is abelian <u>or</u> μ is symmetric (i.e. $\mu(B) = \mu(B^{-1})$). Then there exist elements $a_n \in S$ such that the sequence $\mu^n * \delta_{a_n}$ converges vaguely as $n \to \infty$ to some probability measure iff there exists a compact normal subgroup H such that $S_\mu \subset H \cdot x$ for some $x \notin H$. Also, the limiting measure Q (when it exists) is the translate of some Haar measure on a compact group. [It follows that there do not exist such elements a_n when μ is symmetric.]

Proof. The proof of the 'if' part follows from that of Theorem 4.31. We will prove only the "only if" part.

First, suppose that S is abelian. If $a_n \in S$ such that the sequence $\mu^n * \delta_{a_n}$ converges to some $Q \in P(S)$ as $n \to \infty$, then the sequence

$$\upsilon_n = (\mu^n * \delta_{a_n}) * (\delta_{a_n^{-1}} * \bar{\mu}^n) = \mu^n * \bar{\mu}^n$$

converges, as $n \to \infty$, to $Q * \bar{Q} = \lambda$, say, where for any measure υ, $\bar{\upsilon}$ denotes the measure $\bar{\upsilon}(B) = \upsilon(B^{-1})$. This means that $\upsilon_n = (\mu * \bar{\mu})^n$ and therefore, since $\lambda = \lim_{n\to\infty} (\mu * \bar{\mu})^n$, $\lambda = \lambda^2$ is the normed Haar measure on its support $H(=S_\lambda)$, a compact subgroup. Also, $\mu * \bar{\mu} * \lambda = \lambda$ and therefore, $S_\mu \cdot S_\mu^{-1} \subset H$ and $H = \overline{\bigcup_{n=1}^{\infty} (S_\mu \cdot S_\mu^{-1})^n}$. It is clear that if $x \in S_\mu \cup S_\mu^{-1}$, $x^{-1}HX \subset H$. It follows that for all $x \in S$, $x^{-1}Hx \subset H$ and consequently, H is a compact normal subgroup and $S_\mu \subset H\cdot x$ for some $x \notin H$.

In case, S is not abelian and $\mu = \bar{\mu}$, the sequence υ_n equals μ^{2n} and $\lambda = \lim_{n\to\infty} \mu^{2n}$. Again, $S_\lambda = H$ is a compact subgroup since $\lambda = \lambda^2$. Since $\mu = \bar{\mu}$, $S_\mu = S_\mu^{-1}$. It is clear that $H = \overline{\bigcup_{n=1}^{\infty} S_\mu^{2n}}$ and as above, H is normal and $S_\mu \subset H\cdot x$, $x \notin H$.

Finally, about the structure of the limiting measure Q, (when it exists) let us first consider the case $\mu = \bar{\mu}$. Then $\mu^{2n} \to \lambda$ as $n \to \infty$, where λ is the Haar measure on a compact subgroup H. Since $\mu^{2n} * \delta_{a_{2n}} \to Q$ as $n \to \infty$, it is clear that the subsequence a_{2n} has a cluster point $a \in S$. This means that $Q = \lambda * \delta_a$. Now we consider the case when S is abelian. In this case, with no loss of generality, we assume that $e \in S_Q$. Now by Proposition 4.30, there exists $z \in S$ such that $\mu * Q = Q * \delta_z$ and therefore, by the abelian property of S,

$(\delta_z^{-1} * \mu) * Q = Q$. It follows that $Q(Bx^{-1}) = Q(B)$ for all Borel sets B and all x in H_1, the group generated by the support $z^{-1} \cdot S_\mu$ of the measure $\delta_{z^{-1}} * \mu$. Since $Q * \bar{Q} = \lambda$ and $(\mu * \bar{\mu})^n \to \lambda$, it is clear that $S_Q \subset S_\lambda$ and $S_\lambda = \overline{\bigcup_{n=1}^{\infty} (S_\mu \cdot S_\mu^{-1})^n}$. It follows that $S_Q \subset S_\lambda \subset$ the group generated by $z^{-1} S_\mu$. This means that Q is the Haar measure on S_Q.

Q.E.D.

We remark that if there exist two sequences of elements (a_n) and (b_n) in S such that for $\mu \in P(S)$, the sequence $\mu^n * \delta_{a_n} \to Q_a \in P(S)$ and the sequence $\mu^n * \delta_{b_n} \to Q_b \in P(S)$, then Q_a and Q_b are translates of each other. This is because there are compact sets K_1 and K_2 such that for $n > N$ (some positive integer), $\mu^n * \delta_{a_n}(K_1) > \frac{1}{2}$ and $\mu^n * \delta_{b_n}(K_2) > \frac{1}{2}$, this means that $K_1 a_n^{-1} \cap K_2 b_n^{-1} \neq \phi \ \forall\, n > N$ or $a_n^{-1} b_n \in K_1^{-1} K_2 \ \forall\, n > N$. Therefore, the sequence $a_n^{-1} b_n$ has a cluster point $Z \in S$ and it follows that

$$Q_a = \lim_{n\to\infty} \mu^n * \delta_{a_n} = \lim_{n\to\infty} \mu^n * \delta_{b_n} * \delta_{b_n^{-1} a_n} = Q_b * \delta_{z^{-1}}.$$

We also observe that if S has a quotient (by a compact kernel) which is topologically isomorphic to a group which is abelian, or more generally, a group satisfying the same property as in Theorem 4.31, then for $\mu \in P(S)$ and $S = \overline{\bigcup_{n=1}^{\infty} S_\mu^n}$, the following is true: <u>there exist elements</u> $a_n \in S$ <u>such that the sequence</u> $\mu^n * \delta_{a_n}$ <u>converges to some</u> Q in P(S) <u>iff there exists a compact normal subgroup</u> H <u>such that</u> $S \subset H \cdot x$, $x \notin H$; <u>moreover</u>, <u>the limiting probability measure</u> Q, <u>when it exists</u>, <u>is the translate of a normed Haar measure on a compact subgroup</u>. For instance, let S be connected and maximally almost periodic. Then by [22], the topological commutator subgroup S' is compact and the quotient group (with usual quotient topology) $\frac{S}{S'}$, is a non-compact abelian group. This means that the <u>above result holds in any connected maximally almost periodic group</u>.

Our next result will also shed some light to the question of the validity of the result discussed above.

4.34 <u>Theorem</u>. Let S be <u>nilpotent</u>, $\mu \in P(S)$ and $S = \overline{\bigcup_{n=1}^{\infty} (S_\mu \cup S_\mu^{-1})^n}$. Suppose $e \in S_\mu$. Then for any compact set K,

$$\sup\{\mu^n(Kx) : x \in S\} \to 0$$

as $n \to \infty$.

<u>Proof</u>. Since S is nilpotent, by definition there is a finite sequence of closed normal subgroups $(Z_i)_{i=1}^{n}$ such that

$$\{e\} = Z_o \subset Z_1 \subset Z_2 \subset \ldots \subset Z_n = S$$

and the quotient group Z_{i+1}/Z_i is the center of the quotient group S/Z_i for $i = 0,1,\ldots,n-1$. We make an induction argument on n. If $n = 1$, S is non-compact abelian and the theorem follows by Csiszar's Theorem and Theorem 4.33. Suppose the theorem is true for all non-compact nilpotent groups S whose central ascending series (as above) has length less than n. It is clear that the quotient group S/Z (where Z = the center of S) has length n-1 for its central ascending series, if the corresponding series for S has length n. If S/Z is compact, then by [22], the quotient S/S' (S' = the topological commutator subgroup) is non-compact and abelian with S' compact; therefore, by our remark just preceding this theorem and Csiszar's Theorem, the conclusion of the theorem follows for S. Now the only case left to be considered is when the quotient S/Z is non-compact. Then by induction-hypothesis, the conclusion of the theorem holds for the group S/Z, which is non-compact, nilpotent and has length n-1 for its central ascending series. Let us define the measure λ on the Borel subsets of S/Z by $\lambda(B) = \mu(\phi^{-1}(B))$, where ϕ is the natural map from S onto S/Z. Then for any compact set $K \subset S$, $\phi(K)$ is compact and

$$\sup_{x \in S} \lambda^k(\phi(K)\phi(x)^{-1}) \to 0$$

as $k \to \infty$, since S_λ contains the identity of S/Z. Since for each k,

$$\lambda^k(B) = \mu^k(\phi^{-1}(B)),$$

it follows that the theorem holds for S. The induction argument is complete. The theorem now follows.

Our next theorem in this section gives a complete picture of when there exist elements a_n such that $\mu^n * \delta_{a_n}$ converges weakly in the case of non-compact abelian groups.

4.35 Theorem. Let S be a locally compact non-compact abelian group. Let $\mu \in P(S)$ and $S = \overline{\bigcup_{n=1}^{\infty} (S_\mu \cup S_\mu^{-1})^n}$. Then there exist elements $a_n \in S$ such that the sequence $\mu^n * \delta_{a_n}$ converges vaguely as $n \to \infty$ to a probability measure if and only if the following conditions hold:

(i) S is topologically isomorphic to $Z \times H_o$, where Z is the discrete group of integers and H_o, a compact abelian group;

(ii) $S_\mu = \{1\} \times A$, where A is some compact subset of H such that $\overline{\bigcup_{n=1}^{\infty} (A \cup A^{-1})^n} = H$.

Proof. Suppose there exist $a_n \in S$ such that $\mu^n * \delta_{a_n} \to Q \in P(S)$ as $n \to \infty$. Then by Theorem 4.33, $S_\mu \subset Hx$, $x \notin H$ and H is a compact subgroup. Hence S is compact and consequently, S is compactly generated. By [24, p.90], S is topologically isomorphic to the direct product $R^n \times Z^m \times H$, where R is the additive group of reals, Z is the additive group of integers and H_o is a compact abelian group, and n,m are non-negative integers. If m and n are both positive, then since $S_\mu \subset Hx$ and H is a compact subgroup of S, it is clear after identifying S with $R^n \times Z^m \times H_o$ that

$$S_\mu \subset \{x_1\} \times \{x_2\} \times H_o$$

where $x = (x_1, x_2, x_3) \in S$. But if $n > 0$, then S_μ cannot generate the group S and consequently, $n = 0$. If $m > 1$, then also S_μ cannot generate the group S since a single element cannot generate Z^m, $m > 1$. Hence, $m = 1$ and S is topologically isomorphic with $Z \times H_o$. Since for S_μ to generate $Z \times H_o$, x_2 above must be the integer 1, it is clear that $S_\mu = \{1\} \times A$, $A \subset H_o$.

The converse is clear by Theorem 4.33.

Q.E.D.

We now repeat the following conjecture [48].

4.36 <u>CONJECTURE</u>. Let S be a locally compact non-compact connected group and $\mu \in P(S)$ such that $S = \overline{\bigcup_{n=1}^{\infty} (S_\mu \cup S_\mu^{-1})^n}$. Then for every compact set K,

$$\sup\{\mu^n(Kx) : x \in S\} \to 0$$

as $n \to \infty$ (or equivalently, there don't exist elements a_n such that the sequence $\mu^n * \delta_{a_n}$ is weakly convergent).

Finally in this section, we consider the question of how fast the convolution iterates μ^n of a probability measure converge. In this context, we will restrict ourselves to compact groups and discuss the question of speed of convergence only in norm (the usual variation norm for finite signed measures). For related questions on weak convergence, we refer the reader to [3].

4.37 Theorem. Let μ be a probability measure on a compact Hausdorff group S. If for some positive integer k there exists a number c, $o < c \leq 1$, such that for all Borel sets B,

$$\mu^k(B) \geq c\, m\,(B)$$

where m is the normed Haar measure of S, then

$$||\mu^n - m|| \leq (1-c)^{[n/k]},$$

[n/k] denoting the integer part of n/k.

Proof. Let us write: $\mu^k = \lambda$.

Then $\lambda'(E) = \lambda(E) - m(E)$ defined a finite signed measure on S, and $\lambda' * m = m * \lambda' = 0$. It follows that $\lambda^n = \lambda'^n + m$ for all positive integers n. We define: $\beta(E) = \lambda'(E)+(1-c)m(E)$ $= \lambda(E) - cm(E)$. Also, $\beta^n = \lambda'^n + (1-c)^n \cdot m$. Since $\beta(E) \geq 0$, $\beta^n(E) \geq 0\ \forall\ E \subset S$.

Now we have:

$$\begin{aligned} \lambda^n(E) &= m(E) + \lambda'^n(E) \\ &= [1 - (1-c)^n]\, m(E) + \beta^n(E) \\ &\geq [1 - (1-c)^n]\, m(E). \end{aligned} \tag{59}$$

Replacing E by E^c, we have:

$$1 - \lambda^n(E) \geq [1 - (1-c)^n]\,[1 - m(E)] \text{ or}$$

$$\lambda^n(E) \leq (1-c)^n + [1 - (1-c)^n]\, m(E) \tag{60}$$

Hence from (59) and (60), $\forall\ E \subset S$,

$$|\lambda^n(E) - m(E)| \leq (1-c)^n$$

The rest of the theorem is clear now.

4.38 Corollary. If $\mu \in P(S)$ and S is a finite group, and the sequence μ^n converge weakly to m (the normed Haar measure on S), then $||\mu^n - m||$ converges to zero exponentially fast, as $n \to \infty$.

The proof of 4.38 is immediate from 4.37.

4.39 <u>Theorem</u>. Let S be a compact Hausdorff group and $\mu \in P(S)$ such that for some positive integer k, the absolutely continuous component (with respect to m = the normed Haar measure on S) of μ^k has a support whose m - measure exceeds half. Then $||\mu^n - m||$ converges to zero exponentially fast, as $n \to \infty$.

<u>Proof</u>. It is sufficient to prove the theorem for k = 1. Let μ_o be the absolutely continuous component of μ and f be the density $\frac{d\mu o}{dm}$. It is clear that

$$1 = \int_{S_{\mu_o}} f(x)\ m(dx)$$

$$= \lim_{n \to \infty} \int_{S_{\mu_o} \cap A_n} f(x)\ m(dx)$$

where $A_n = \{x \in S : f(x) \geq \frac{1}{n}\}$. This means that we can choose a positive number c such that if $A = \{x \in S : f(x) \geq c\}$, then $m(A) > \frac{1}{2}$. Now

$$\begin{aligned} ||\mu - cm|| &\leq ||\mu - \mu_o|| + ||\mu_o - cm|| \\ &= (\mu - \mu_o)(S) + \mu_o(A) - cm(A) \\ &\qquad + cm(A^c) - \mu_o(A^c) \\ &\leq (\mu - \mu_o)(S) + \mu_o(A) - cm(A) \\ &\qquad\qquad + c.[1 - m(A)] \\ &< (\mu - \mu_o)(S) + \mu_o(A) \\ &\leq \mu(S) = 1. \end{aligned}$$

Write: $\upsilon = (1-c)^{-1}(\mu - cm)$. Then

$$\begin{aligned} \mu^n &= [(1-c)\ \upsilon + c \cdot m]^n \\ &= \sum_{r=1}^{n-1} \binom{n}{r} (1-c)^r \cdot c^{n-r} \cdot \upsilon^r m^{n-r} \\ &\qquad + c^n m + (1-c)^n \upsilon^n \\ &= \sum_{r=o}^{n-1} \binom{n}{r} (1-c)^r c^{n-r} m + (1-c)^n \upsilon^n + c^n m \end{aligned}$$

$$= m - (1-c)^n m + (1-c)^n \upsilon^n.$$

Hence,

$$||\mu^n - m|| \leq (1-c)^n + (1-c)^n \cdot ||\upsilon||^n.$$

Since $||\upsilon|| < (1-c)^{-1}$, the theorem follows.

Our last theorem in this section is the following.

4.40 Theorem. If the probability measure μ on a compact Hausdorff connected group S has a non-zero absolutely continuous component with respect to the normed Haar measure m on S, then $||\mu^n - m||$ converges to zero exponentially fast as n tends to infinity.

Proof. Let f be the density of the absolutely continuous component of μ. With no loss of generality, we assume that f is bounded and consequently by [24,p.295], μ^2 has a continuous density g. Let V be the set $\{x \in S : g(x) > 0\}$. Then V is open and since S is connected, for some integer p, $V^p = S$. Since $V \subset S_{\mu^2}$, μ^{2p} has an absolutely continuous component whose support has m - measure 1. Hence the theorem follows by 4.39.

Comments on the results of section 4

Propositions 4.3 and 4.4 are due to M. Rosenblatt. The important theorem 4.13 is also due to Rosenblatt, see [66]. Propositions 4.5 - 4.7 as well as 4.7A and 4.7B are taken from Mukherjea [47]. Theorem 4.8 (in its present form) is taken from [68]; the results in this theorem are well-known and due to Kawada-Ito [30], and the part (i) equivalent to (ii) is due to Collins [12]. Theorem 4.12 is due to Kloss [33]. Theorems 4.14, 4.14A, 4.14B, 4.15 and, 4.17, 4.18 and 4.19 are results of Mukherjea and they are taken from [47] and [48]. Theorem 4.20 is taken from Kloss [34]. The results in 4.21A, 4.21B, 4.22 and 4.22 A through E are taken from B. Center and A. Mukherjea [5]. 4.16 is due to B. Center.

Extensions of the classical Levy equivalence theorem are obtained by Csiszár [13] and Galmarino [17] in the case of locally compact groups. More complete results in this context are obtained in the case of finite groups by Maximov [44] and in the case of discrete completely simple semigroups by A. Mukherjea and T. C. Sun. The results in 4.23, 4.24, 4.25, 4.26 and 4.28 are taken from Mukherjea and Sun [51]. Theorems 4.31, 4.33, 4.34 and 4.35 are results of B. Center and A. Mukherjea; they are taken from [5]. The question of the speed of convergence of the convolution iterates is first considered in Kloss [33] and later in Bhattacharya [3]. Theorem 4.37 is essentially given in [33] and Theorems 4.39 and 4.40 are taken from [3].

In the context of some of the above results in discrete semigroups, the paper [41] of Martin-Löf is one of the most interesting papers. His results on the convergence of convolution iterates of a probability measure are complete in the case of discrete semigroups. Because of the restrictive size of these notes, these results along with many other results couldn't be included here.

5. Limit Behavior of Convolution iterates of a probability measure on a topological semigroup: Two concrete examples

In this section, we will study the behavior of the limit (weak*) of the convolution iterates of a probability measure on two special semigroups--first, on the semigroup $[0, \infty)$ of non-negative real numbers under multiplication and usual topology and then, on the semigroup of stochastic matrices.

A. The First Example: The Semigroup $[0, \infty)$

The study and consideration of this example is a result of an attempt to prove or disprove Rosenblatt's result Proposition 4.4 on non-compact topological semigroups. It turns out that his result does not extend to the non-compact situation, as will be evident in what follows.

Let $S = [0, \infty)$ (as described above) and μ be the normalized Lebesgue measure with support $F' = [0, a]$, $a > 1$. Then F generates S. Since $\{0\}$ is the smallest ideal of S, we wish to determine, among other things, the behavior of $\mu^n([0, \alpha])$, $0 < \alpha$.

We claim the following:

(i) If $a < e$, then for $0 < \alpha$, $\mu^n([0, \alpha])$ converges to 1 as $n \to \infty$.

(ii) If $a > e$, then for $0 < \alpha$, $\mu^n([0, \alpha])$ converges to 0 as $n \to \infty$.

(iii) If $a = e$, then for $0 < \alpha$, μ^n $([0, \alpha])$ converges to 1/2 as $n \to \infty$.

(iv) For all $a > 1$, $\mu^n(K)$ converges to 0 as $n \to \infty$, whenever K is compact and doesn't contain 0.

Actually the proof of the first three assertions, which utilizes the Central Limit Theorem of probability theory, shows that it is possible to obtain a more general class of measures (not necessarily Lebesgue measures) for which Rosenblatt's result fails to hold on general locally compact semigroups. After the proofs are given we indicate how the above assertions are related to certain limiting properties of the sections of the Taylor expansion of the exponential function.

<u>Proof</u>; Let $X_1, X_2, \ldots$ be a sequence of independent, identically distributed random variables on some probability space, with values in $[0, \infty)$ and with distribution μ, i.e., $P[X_i \in B] = \mu(B)$, where P is the probability measure and B is any Borel set on $[0, \infty)$.

Let $Z_n = X_1 \cdot X_2 \cdots X_n$. Since the X_i's are independent, $P(Z_n \in B) = \mu^n(B)$. Consider the sequence $\log Z_n$, which is clearly defined with probability 1. Now

$$\log Z_n = \sum_{i=1}^{n} \log X_i = \sum_{i=1}^{n} Y_i,$$

where $Y_i = \log X_i$. We wish to apply the Central Limit Theorem to the sequence $Y_1, Y_2 \cdots$. To do so, we must check if the mean $m = E(Y_i)$ and the variance $\sigma^2 = E(Y_i^{\,2}) - m^2$ are finite.

Clearly,

$$m = \frac{1}{a} \int_0^a \log x \, dx$$

$$= \log a - 1,$$

and

$$\sigma^2 + m^2 = \frac{1}{a} \int_0^a (\log x)^2 \, dx$$

$$= (\log a - 1)^2 + 1,$$

so that both m and σ^2 are finite. Hence by the Central Limit Theorem, the distribution of

$$\frac{\sum_{i=1}^{n} Y_i - n \cdot m}{n^{1/2} \sigma}$$

converge to N(0, 1), the normal distribution with mean 0 and variance 1, as $n \to \infty$. We wish to find $\lim_{n\to\infty}\mu^n([0, \alpha])$, i.e., the limit of $P(Z_n \varepsilon [0, \alpha])$ as $n \to \infty$. To do this, we have to find sets $A_n \subset (-\infty, \infty)$ such that the following set equation holds

$$n^{1/2}\sigma A_n + n\cdot m = \log (0, \alpha),$$

i.e.

(a) $$A_n = (-\infty, n^{1/2}\cdot \frac{-m}{\sigma} + \frac{1}{n^{1/2}\sigma} \log \alpha).$$

Case (i): $a < e$. In this case, $m = \log a - 1 < 0$. Since $-m > 0$, it is clear from (a) that given any positive integer p, we can find N_p such that $n > N_p$ implies $A_n \supset (-\infty, p)$. Then, we have for $n > N_p$,

$$\mu^n([0, \alpha]) = P(Z_n \varepsilon [0, \alpha]) = P(\log Z_n \varepsilon \log (0, \alpha))$$

$$= P\left[\frac{\sum_{i=1}^{n} Y_i - n\cdot m}{n^{1/2}\sigma} \varepsilon A_n\right]$$

$$\geq P\left[\frac{\sum_{i=1}^{n} Y_i - n\cdot m}{n^{1/2}\sigma} \varepsilon (-\infty, p)\right],$$

where the last term converges, as $n \to \infty$, to

$$\frac{1}{(2\pi)^{1/2}} \int_{-\infty}^{p} e^{-x^2} dx,$$

by the Central Limit Theorem. Since

$$\lim_{p\to\infty} \frac{1}{(2\pi)^{1/2}} \int_{-\infty}^{p} e^{-x^2} dx = 1,$$

it follows that $\lim_{n\to\infty} \mu^n([0, \alpha]) = 1$.

Case (ii): $a > e$. In this case, $m = \log a - 1 > 0$. Since $m > 0$, we see from above that given any positive integer p, there exists an N_p such that $n > N_p$ implies $A_n \subset (-\infty, -p)$. Now for $n > N_p$, we have

$$\mu^n([0,\alpha]) = P\left(\frac{\sum_{i=1}^{n} Y_i - n\cdot m}{n^{1/2}\sigma} \varepsilon A_n\right) \leq P\left(\frac{\sum_{i=1}^{n} Y_i - n\cdot m}{n^{1/2}\sigma} \varepsilon (-\infty, -p)\right)$$

which converges, as $n \to \infty$, to

$$\frac{1}{(2\pi)^{1/2}} \int_{-\infty}^{-p} e^{-x^2} dx.$$

Since

$$\int_{-\infty}^{-p} e^{-x^2} dx \to 0 \text{ as } p \to \infty,$$

it is clear that $\lim_{n\to\infty} \mu^n([0, \alpha]) = 0$.

Case (iii): $a = e$. In this case, $m = \log a - 1 = 0$. Therefore,

$$A_n = \left(-\infty, \frac{1}{n^{1/2}\sigma} \log \alpha\right).$$

Since, as $n \to \infty$, $(\log \alpha)/n^{1/2} \sigma \to 0$, and

$$\frac{1}{(2\pi)^{1/2}} \int_{-\infty}^{0} e^{-x^2} dx = 1/2,$$

by similar arguments as in previous cases, it follows easily that

$$\lim_{n\to\infty} \mu^n([0, \alpha]) = 1/2.$$

[<u>Note</u>. Clearly, from the case $a > e$, one can see that if we take any probability measure (not necessarily a Lebesgue measure) whose support generates $[0, \infty)$ such that $m > 0$ and the mean and the variance of Y_i are finite, then $\lim_{n\to\infty} \lambda^n[0, \alpha] = 0$ for every $\alpha > 0$.]

We now remark that the assertions (i)-(iii) are actually equivalent to certain statements concerning the sections

$$S_n(x) = \sum_{i=0}^{n} \frac{x^i}{j!}$$

of the Taylor expansion of e^x. To see this we first notice that for $0 < \alpha \leq a^n$, $n > 2$, we have

$$\mu^n([0, \alpha]) = \frac{1}{a} \int_0^a \mu^{n-1}([0, \alpha/x])\, dx$$

$$= \frac{1}{a} \int_0^r dx + \frac{1}{a} \int_r^a \mu^{n-1}([0, \alpha/x])\, dx,$$

where $r \equiv \alpha/a^{n-1}$. By using induction it then follows that

$$\mu^n([0, \alpha]) = \frac{\alpha}{a^n} \sum_{j=0}^{n-1} \left(\log \frac{a^n}{\alpha}\right)^j \cdot \frac{1}{j!} .$$

On setting $b \equiv \log a$ (>0) and $\gamma \equiv -\log \alpha$ this last equation becomes

$$\mu^n([0, \alpha]) = \frac{1}{e^{nb+\gamma}} \sum_{j=0}^{n-1} (nb + \gamma)^j \cdot \frac{1}{j!}$$

$$= \frac{1}{e^{nb+\gamma}} [S_n(nb + \gamma) - \frac{(nb + \gamma)^n}{n!}].$$

Now by applying Stirling's formula and the fact that $eb/e^b \leq 1$, it is easy to see that

$$\lim_{n\to\infty} \frac{(nb + \gamma)^n}{n!\, e^{nb+\gamma}} = 0,$$

and so the assertions (i), (ii), and (iii) are equivalent to the conditions that

$$\lim_{n\to\infty} \frac{S_n(nb + \gamma)}{e^{nb+\gamma}} = \begin{cases} 1, & \text{if } 0 < b < 1, \\ 0, & \text{if } b > 1, \\ 1/2, & \text{if } b = 1, \end{cases}$$

for each fixed $\gamma \geq 0$.

B. The Second Example: The Semigroup Of 2 X 2 Stochastic Matrices.

In this example, we'll study the behavior of the limit of the convolution iterates of a probability measure on the compact semigroup of 2x2 stochastic matrices. For simplicity (and for reasons of mathematical difficulties), we'll only consider the iterates of a probability measure whose support consists of two points only.

Let S be the semigroup of all 2x2 stochastic matrices (all entries are non-negative and the row sums equal one). There is then a natural one-to-one mapping from S onto the unit square, namely the one which maps each matrix to the point whose co-ordinates are the entries of its first column. Thus giving S the topology of the unit square, S becomes a compact topological semigroup with usual matrix multiplication. The kernel K of S is the set of all 2x2 stochastic matrices with identical rows and is topologically isomorphic to the main diagonal of the unit square (the one that joins (0,0) and (1,1)). This kernel is a right-zero semigroup. Let μ be a probability measure on S whose support S_μ contains a matrix with non-zero entries. Then the closed semigroup D generated by S_μ intersects K and $D \cap K$, a right-zero semigroup, is the kernel of D. By Theorem 4.13, the sequence μ^n converges weakly to a probability measure υ whose support is $D \cap K$. This measure satisfies the convolution equation

$$(1)\cdots \qquad \upsilon * \mu = \upsilon$$

In fact, the limit υ of μ^n is the <u>unique</u> solution of the equation (1). For, let β be another solution of (1) and $\beta * \mu = \beta$, with $S_\beta \subset D$. Then by Prop. 4.5, $\beta = \beta^2$ and $S_\beta \subset D \cap K$. Since $\beta * \mu^n = \beta$ for each positive integer n and $\mu^n \to \upsilon$ as $n \to \infty$, it is clear that $\beta * \upsilon = \beta$. But β and υ are both in $P(D \cap K)$ and $D \cap K$ is a right-zero semigroup, meaning that $\beta * \upsilon = \upsilon$. Hence $\beta = \upsilon$, proving that the solution of (1) is unique.

Suppose now that $S\mu$ is a two-point set $\{(x_1,y_1), (x_2,y_2)\}$ such that $\mu(\{(x_1,y_1)\}) = p$ and $\mu(\{(x_2,y_2)\}) = q$ and that

$G(x) = \upsilon(\{(t, t) : t \leq x\})$.

Then the equation (1) is equivalent to

(2) $G(x) = p\, G\left(\frac{x - y_1}{x_1 - y_1}\right) + q\, G\left(\frac{x - y_2}{x_2 - y_2}\right)$.

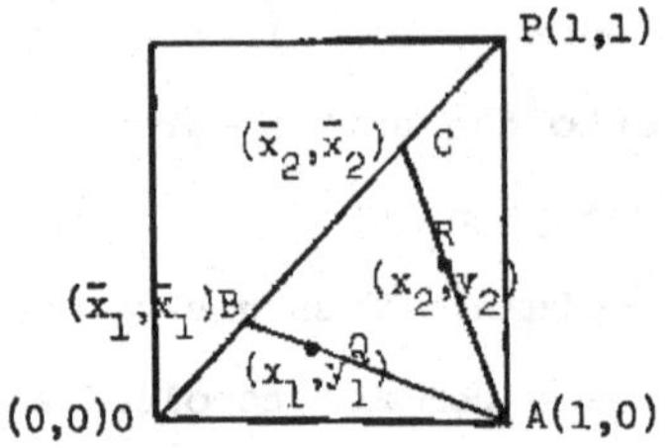

It is clear that the semigroup generated by the support of μ is inside the triangle ABC (including the boundary).

[Note that if μ were the unit mass at (x_1,y_1), then μ^n would converge to the unit mass at B.] Since $S_\upsilon \subset$ the segment BC, G(x) is zero on [O, B] and one on [C, P].

5.2 __Proposition__. The function G(x) is continuous iff the points (x_1,y_1), (x_2,y_2) and (0,1) are not collinear.

__Proof__. If the points (x_1,y_1), (x_2,y_2) and (0,1) are collinear, then υ has one point support and consequently, G is __not__ continuous. If these points are not collinear, then $B \neq C$ (see the diagram above). To prove that G is continuous, it is sufficient to prove that $\upsilon(\{t\}) = 0$ for each t in [B,C]. Now we suppose that

$$\sup \{\upsilon(\{t\}) : t \in [B,C]\} = s > 0.$$

Then if this supremum is not attained, there will exist t_n such that $\upsilon(\{t_n\}) > \frac{s}{2}$ for all n and the t_n's are all different. But this is impossible since $\upsilon(S) = 1$. Therefore, there must be a point t such that $\upsilon(\{t\}) = s$. Since $\upsilon * \mu = \upsilon$, then we have:

$$\upsilon(\{t\}) = p\,\upsilon(t\,\bar{Q}^1) + q\upsilon(t\,\bar{R}^1)$$

where Q, R are the points of S_μ. It is clear that then, $\upsilon(t\bar{Q}^1) = \upsilon(t\bar{R}^1) = s$. Since $B \neq C$, the point t is different from either B or C. Suppose $t \neq B$. Then for all n, the points $t\bar{Q}^n$ are all different, and this contradicts the finiteness of υ.

The proposition now follows easily.

5.3 Proposition. The function $G(x)$ is either singular or absolutely continuous.

Proof. Suppose G is not singular. Then $G'(x)$ (the derivative of G) is positive for all x in some set M of positive Lebesgue measure. Let us define:

$$F(t) = [\int_0^t G'(x)\, dx]/[\int_0^1 G'(x) dx].$$

Then F is a absolutely continuous mapping from $[0,1]$ onto $[0,1]$. It can be verified easily that if β is the probability measure induced by F on $[0,1]$, then $\beta * \mu = \beta$. By the uniqueness of the solution of this equation, $\beta = \upsilon$ and consequently, $G = F$.

Q.E.D.

5.4 Proposition. If $H(x)$ is an increasing function that satisfies (2) and if $H(0) = 0$, $H(1) = 1$, then $H(x) = G(x)$.

Proof. It is clear that the function $H(x+) = \lim_{y \to x+} H(y)$ also satisfies (2). Since $H(x+)$ is a distribution function, by the uniqueness of the solution of (2), $H(x+) = G(x)$. Since $G(x)$ is continuous and the function $H(x)$ can have at the most countably many discontinuities, $G(x) = H(x)$.

Q.E.D.

Now we wish to write equation (2) in a more convenient form. We write:

$$x_o = \frac{y_1}{1-(x_1-y_1)}, \quad L = x_o - \frac{x_o - y_2}{x_2 - y_2}.$$

Then $x_o = \dfrac{x_o - y_1}{x_1 - y_1}$ and writing $g(x) = G(Lx + xo)$, we have from equation (2),

$$g(x) = p\,G\left(\frac{Lx + x_o - y_1}{x_1 - y_1}\right) + q\,G\left(\frac{Lx + x_o - y_2}{x_2 - y_2}\right)$$

$$= p\,G\left(\frac{Lx}{x_1 - y_1} + x_o\right) + q\,G\left(\left(\frac{x}{x_2 - y_2} - 1\right)L + x_o\right)$$

$$(3)\cdots \text{ or } g(x) = p\,g\left(\frac{x}{a}\right) + q\,g\left(\frac{x}{b} - 1\right)$$

where $a = x_1 - y_1$ and $b = x_2 - y_2$.

Now we intend to find the solution of the equation (3). For any real number x, write

$$x\,T_1 = \frac{x}{a} \quad\text{and}\quad x\,T_2 = \frac{x}{b} - 1.$$

Then let W be the free semigroup, with unit I, generated by T_1 and T_2. Let us write: $t \downarrow s$ if t is a proper left divisior of s in W. We now define:

(i) $V(t) = p^k q^m$,

where t is the product of k many T_1's and m many T_2's; and

(ii) $h(x) = \sum_{s \in A(x)} V(s)$,

where $A(x) = \{s \in W : xs > \frac{b}{1-b} \text{ and } t \downarrow s \Rightarrow xt \leq \frac{b}{1-b}\}$.

Then we observe that

(i) The function h(x) is an increasing function. To see this, let $x < x'$ and $s \in A(x)$. Then either $s \in A(x')$ or there is a proper divisor t of s such that $t \in A(x')$. But since $V(t) \geq \sum_{\substack{t \downarrow s \\ s \in A(x)}} V(s)$ (this can be readily verified by the definition of V), it is clear that h is increasing.

(ii) The function h(x) satisfies the equation (3). To see this, we notice that if $A(x) = \{I\}$, then $A(x) = A\left(\frac{x}{a}\right) = A\left(\frac{x}{b} - 1\right)$ and if $A(x) = \phi$, then $A(x) = A\left(\frac{x}{a}\right) = A\left(\frac{x}{b} - 1\right)$. In all other cases,

$$A(x) = T_1\,A\left(\frac{x}{a}\right) \cup T_2\,A\left(\frac{x}{b} - 1\right).$$

This implies that $h(x) = p\,h\,(\frac{x}{a}) + q\,h\,(\frac{x}{b} - 1)$ and therefore, h is a solution of (3).

In what follows, the nature of the solution of the equation (2) or (3) will be studied. More specifically, we wish to determine conditions on the support of μ which will force the limit υ to be absolutely continuous or continuous singular.

5.5 Proposition. Suppose $(x_1 - y_1) + (x_2 - y_2) < 1$. Then the function G(x) is continuous singular. [Here we assume that $B \neq C$.]

Proof. As in the diagram, let $\bar{x}_2 > \bar{x}_1$, where $C = (\bar{x}_2, \bar{x}_2)$ and $B = (\bar{x}_1, \bar{x}_1)$. Let $w = BR$ and $z = C \cdot Q$. [Recall that B, R, C and Q all represent certain stochastic matrices] By direct calculations, $w > z$ iff $(x_1 - y_1) + (x_2 - y_2) < 1$. Also, it follows directly from the functional equation (2) that $G(w) = G(z) = p$ and therefore, G is constant on the interval $I_o = [z, w]$. The length $\ell(I_o)$ of I_o is $(1 - (x_1 - y_1) - (x_2 - y_2))(\bar{x}_2 - \bar{x}_1)$.

Let us consider

$$A = \{I_o x : x \in DR\}$$

and $$B = \{I_o x : x \in DQ\}$$

where D is the semigroup generated by Q and R. Then A and B are each a countable collection of disjoint open intervals and

$$\bigcup A \subset (w, C) \text{ and } \bigcup B \subset (B, z).$$

It follows from (2) that G is constant on each interval in A or B. Now the length of I_oR is $\ell(I_o) \cdot (x_2 - y_2)$ and the length of I_oQ is $\ell(I_o)\ (x_1 - y_1)$. Hence, by direct computations (after simplifications), we have: if m is the Lebesgue measure, then

$$m(UA) = \frac{\ell(I_o) \cdot (x_2 - y_2)}{1-(x_1 - y_1)-(x_2 - y_2)}$$

and $m(UB) = \dfrac{\ell(I_o)\cdot(x_1 - y_1)}{1-(x_1 - y_1) - (x_2 - y_2)}$

Hence, $m\ (I_o \cup (UA) \cup (UB)) = \dfrac{\ell(I_o)}{1-(x_1 - y_1) - (x_2 - y_2)}$ = the length of the interval [B, C]. This means that G'(x) is zero almost everywhere on [B, C] and therefore, G(x) is singular.

Q.E.D.

5.6 Proposition. If $p = x_1 - y_1$ and $q = x_2 - y_2$, then G(x) is absolutely continuous.

Proof. In this case, it can be verified directly that

$$\begin{aligned} G(x) &= 0, \quad 0 \leq x \leq \bar{x}_1 \\ &= [x - \bar{x}_1]/[\bar{x}_2 - \bar{x}_1], \quad \bar{x}_1 \leq x \leq \bar{x}_2 \\ &= 1, \quad \bar{x}_2 \leq x \leq 1. \end{aligned}$$

The rest is clear.

Q.E.D.

5.7 Proposition. Suppose that $(x_1 - y_1) + (x_2 - y_2) = 1$. Then the following are true.

(i) For every x in $(0, \frac{b}{a})$, there is a unique representation of x as $\sum_{k=1}^{\infty} b^k a^{m_k}$, where $m_k \leq m_{k+1}$ and m_k's are zero or natural numbers;

(ii) If $x = \sum_{k=1}^{\infty} b^k a^{m_k}$, then the solution g of the equation (3) has its value at x given by : $g(x)=\frac{p}{q} \sum_{k=1}^{\infty} q^k p^{m_k}$.

(iii) If $p \neq x_1 - y_1$, then the function g(x) is continuous singular. [Recall: $a = x_1 - y_1$ and $b = x_2 - y_2$.]

Proof of (i). For each positive integer n, let m_n be the smallest non-negative integer such that $m_n \leq m_{n+1}$ and $\sum_{k=1}^{n} b^k a^{m_k} \leq x$. If for some n, the equality is attained here, then the representation in (i) follows. Otherwise, there exists N such that for n > N, we have

$$\sum_{k=1}^{n} b^k a^{m_k} < x \le \sum_{k=1}^{n-1} b^k a^{m_k} + b^n a^{m_n - 1};$$

this means that

$$x - \sum_{k=1}^{n} b^k a^{m_k} \le b^{n+1} a^{m_n - 1},$$

which goes to zero as $n \to \infty$. This proves that $x = \sum_{k=1}^{\infty} b^k a^{m_k}$, as claimed in (i). To prove the uniqueness part, for x in $(0, \frac{b}{a})$, let

$$x = \sum_{k=1}^{\infty} b^k a^{m_k} = \sum_{k=1}^{\infty} b^k a^{n_k}$$

where $m_k \le m_{k+1}$, $n_k \le n_{k+1}$, and the m_k's and n_k's are non-negative integers. Let j be the first integer such that $m_j \ne n_j$. Let $n_j > m_j$. Then since

$$b^j a^{m_j} = \sum_{k=j}^{\infty} b^k a^{m_j + 1}$$

$$\ge \sum_{k=j}^{\infty} b^k a^{n_k},$$

we have: $\sum_{k=1}^{\infty} b^k a^{m_k} > \sum_{k=1}^{\infty} b^k a^{n_k}$, a contradiction. The proof of (i) is now complete.

<u>Proof of (ii)</u>. Let the function f be defined by:

$$f(x) = 0, \quad x \le 0$$

$$= 1, \quad x \ge \frac{b}{a}$$

$$= \frac{p}{q} \sum_{k=1}^{\infty} q^k p^{m_k}, \text{ if } x = \sum_{k=1}^{\infty} b^k a^{m_k}.$$

Then it is clear that f is a monotonic increasing function. Also f satisfies the equation (3). To see this, let $m_1 = 0$. Then $\frac{x}{b} - 1 \ge 0$ and $\frac{x}{a} \ge \frac{b}{a}$. Since $\frac{x}{b} - 1 = \sum_{k=1}^{\infty} b^k a^{m_{k+1}}$, we have

$$f(x) = \frac{p}{q} \sum_{k=1}^{\infty} q^k p^{m_k}, \quad f\left(\frac{x}{b} - 1\right) = \frac{p}{q} \sum_{k=1}^{\infty} q^k p^{m_{k+1}}.$$

It is clear that $f(x) = pf\left(\frac{x}{a}\right) + qf\left(\frac{x}{b} - 1\right)$.

Let $m_1 \ge 1$. Then $\frac{x}{b} - 1 \le 0$ and $\frac{x}{a} = \sum_{k=1}^{\infty} b^k a^{m_k - 1}$. It is clear that then $f(x) = \frac{p}{q} \sum_{k=1}^{\infty} q^k p^{m_k} = pf\left(\frac{x}{a}\right)$. The proof of (ii) now follows from Prop. 5.4.

Proof of (iii). Suppose that $q > b$. [Note that if $p \neq a$, then $q \neq b$.] Let λ be a real number such that $b > \lambda > q$. For positive integers n, let us consider all the intervals of the form

$$\left(\sum_{j=1}^{k} b^j a^{m_j}, \sum_{j=1}^{k} b^j a^{m_j} + b^{k+1} a^{m_k} \right)$$

where $0 \leq m_1 \leq m_2 \leq \cdots \leq m_k = m$, $m+k-1 = n$ and $k-1 \leq \lambda n$. These intervals can be easily seen to be pairwise disjoint. Also we observe that there are $\binom{m+k-1}{k-1}$ different intervals of this form. The total length of all these intervals is:

$$\sum_{k=1}^{[\lambda n]+1} \binom{n}{k-1} b^{k+1} a^{n+1-k}$$

$$= b^2 \sum_{k=0}^{[\lambda n]} \binom{n}{k} \cdot b^k a^{n-k}$$

which can be approximated by

$$b^2 \Phi \left(\frac{(\lambda-b)n}{\sqrt{nab}}\right), \text{ where}$$

Φ is the normal distribution, and this tends to zero as $n \to \infty$ since $\lambda - b < 0$. On the other hand, the sum of the υ-measures of these intervals is:

$$\sum_{k=1}^{[\lambda n]+1} \binom{n}{k-1} \frac{p}{q} \cdot q^{k+1} p^{n+1-k}$$

$$= p.q \sum_{k=0}^{[\lambda n]} \binom{n}{k} q^k p^{n-k}$$

which can be approximated by

$$p \cdot q \, \Phi \left(\frac{(\lambda-q)n}{\sqrt{npq}}\right)$$

which converges to $p \cdot q$ as $n \to \infty$ since $\lambda > q$. It follows that υ is not absolutely continuous with respect to the Lebesgue measure and the proof of (iii) is complete.

We now consider the case: $a + b > 1$. In what follows, we'll see that in this case, $g(x)$ may or may not be absolutely continuous.

5.8 Proposition. If $p = q = \frac{1}{2}$ and $a = b = \frac{1}{\sqrt{2}}$, then $g(x)$ is absolutely continuous.

Proof. We define the function $h(x)$ as follows:

$$\begin{aligned} h(x) &= 0, \ x \notin (0, \sqrt{2}+1); \\ &= 1, \ x \in (0, 1) = I_1; \\ &= 0, \ x \in [1, \sqrt{2}] = I_2; \\ &= -1, \ x \in (\sqrt{2}, \sqrt{2}+1) = I_3. \end{aligned}$$

Then $h(x) = h(\sqrt{2}\, x) + h\,(\sqrt{2}\, x-1)$ a.e.

Now let us define:

$$h_1(x) = \int_{-\infty}^{x} h(t)\, dt \text{ and } h_2(x) = \int_{-\infty}^{x} h_1(t)\, dt.$$

Then $h_2(x)$ is absolutely continuous and monotonic increasing, and 0 if $x < 0$, constant if $x \geq \sqrt{2} + 1$; also it satisfies:

$$h_2(x) = \frac{1}{2} h_2\,(\sqrt{2}\, x) + \frac{1}{2} h_2\,(\sqrt{2}\, x-1).$$

Hence, $h_2(x)$ (suitably normalized) is $g(x)$. Q.E.D.

We remark that the above result also holds even if $p = q = \frac{1}{2}$ and $a = b = 2^{-\frac{1}{n}}$ for some positive integer n. However, the proof in this general case is quite different and omitted.

Now we will show that the solution $g(x)$ is continuous singular when $a + b > 1$ and $(\frac{p}{a})^p \cdot (\frac{q}{b})^q \geq 1$. To show this, we first find a more convenient expression for the solution of equation (3).

Consider a series

$$(4)\cdots \quad r = \sum_{j=1}^{k} b^j a^{m_j}, \ m_1 \leq m_2 \leq \cdots \leq m_k$$

such that

$$(5)\cdots \quad \sum_{j=1}^{k-1} b^j a^{m_j} \leq x < \sum_{j=1}^{k} b^j a^{m_j}.$$

[Note that the sums of two such series which are different may be equal.] For each x, let $A(x)$ be the set of all such series. For $r \in A(x)$, we define: for r as in (4),

$$V(r) = q^k \; p^{m_k}$$

and

$$(6)\cdots \quad f(x) = \begin{cases} 1, & \text{if } x \leq 0 \\ \sum_{r \in A(x)} V(r), & \text{if } 0 < x < \frac{b}{1-b} \\ 0, & \text{if } x > \frac{b}{1-b}. \end{cases}$$

Note that

$$(7)\cdots \quad A(x) = a\,A\left(\tfrac{x}{a}\right) \cup b\,\left(1 + A\left(\tfrac{x}{b} - 1\right)\right),$$

where the union is clearly disjoint.

5.9 Proposition. If f is the function as given in (6) above, then $g(x) = 1 - f(x)$ is the solution of equation (3).

Proof. Because of (7), it is easy to verify that $f(x) = pf\left(\frac{x}{a}\right) + qf\left(\frac{x}{b} - 1\right)$. Because of Prop. 5.4, it is sufficient to prove that f is monotonically decreasing. To prove this, let $x < x'$. Let $r \in A(x)$. Then either $r \in A(x')$ or there are extensions of r in $A(x')$ or no extension of r is in $A(x')$. [We call a series r' of (k+m) terms whose first k terms are identical to those of r an extension or a m-term extension of r.] We claim that

$$(8)\cdots \quad V(r) \geq \sum_{r' \in A(r,x')} V(r')$$

where $A(r,x') = \{r' : r' \text{ is an extension of } r \text{ and } r' \in A(x')\}$.

To prove our claim, let $r = \sum_{j=1}^{k} b^j a^{m_j}$ and A_1 be a set of one term extensions of r Then

$$\sum_{r' \in A_1} V(r') \leq q^{k+1}\, p^{m_k} + q^{k+1}\, p^{m_k+1} + \cdots$$
$$= q^k\, p^{m_k} = V(r).$$

By induction, we can prove this inequality for any set A_n of n-term or less extensions of r having the property that if r', $r'' \in A_n$, then r" cannot be an extension of r'. This proves (8). Because of (8), it is clear that f(x) is decreasing.

Q.E.D.

5.10 Proposition. The solution $g(x)$ of equation (3) is continuous singular if $a + b > 1$ and $(\frac{p}{a})^p (\frac{q}{b})^q \geq 1$.

Proof. Let us consider the finite series

$$r = \sum_{j=1}^{k} b^j a^{m_j} + b^{k+1} a^{m_k + n}$$

and let $x_1 = \sum_{j=1}^{k} b^j a^{m_j}$, $z_1 = \sum_{j=1}^{k} b^j a^{m_j} + b^{k+1} a^{m_k}$. If n is sufficiently large, $r \in A(x_1)$ and no extension of r (see the underlined sentences above) will be in $A(z_1)$. This is because $\sum_{i=1}^{\infty} b^{k+i} a^{m_k+n} < b^{k+1} a^{m_k}$ if $a^n + b < 1$. Now this n is independent of k and using Prop. 5.9,

$$g(z_1) - g(x_1) = f(x_1) - f(z_1)$$
$$\geq q^{k+1} p^{m_k+n} = A\, q^{k-1} p^{m_k}$$

where $A = q^2 p^n$.

Let us consider (x, z), a finite union of intervals $I_i = (x_i, z_i)$ of the form

$$(9)\cdots \qquad I_i = \Big(\sum_{j=1}^{k_i} b^j a^{m_{j_i}},\ \sum_{j=1}^{k_i} b^j a^{m_{j_i}} + b^{k_i+1} a^{m_{(k_i)_i}}\Big)$$

such that (i) for every i, there is a finite series r_i such that r_i is in $A(x_i)$, but r_i is not in $A(z_i)$ and (ii) if $i \neq j$, then $r_i \neq r_j$ and no extension of r_i can extend r_j. Then we claim:

$$(10)\cdots \qquad g(z) - g(x) \geq \sum_i V(r_i)$$

To prove (10), let $r \in A(x)$ and $B(r)$ be the set of all r_i's which are extensions of r. Then if $r \neq r'$, then $B(r) \cap B(r')$ is empty. Moreover,

$$V(r) \geq \Sigma\ \{V(r') : r' \in B(r) \cup A(r,z)\}$$

or $$V(r) - \sum_{r' \in A(r,z)} V(r') \geq \sum_{r' \in B(r)} V(r')$$

Now we have:

$$g(z) - g(x) = \sum_{r \in A(x)} [v(r) - \sum_{r' \in A(r,z)} v(r')]$$

$$\geq \sum_i v(r_i),$$

which proves the claim (10).

Now we recall that we have assumed $(\frac{p}{a})^p \, (\frac{q}{b})^q \geq 1$. Since $a + b > 1$, one of $\frac{p}{a}$ and $\frac{q}{b}$ is less than one and the other one is greater than one. Suppose that $\frac{p}{a} > 1 > \frac{q}{b}$ and let $\lambda = \frac{qa}{pb} < 1$. Consider for every n, the collection F of all finite series $r = \sum_{j=1}^{k} b^j a^{m_j}$ such that

$$k + m_k = n + 1, \; k \leq nq + 1$$

and for each r, the interval

$$(\sum_{j=1}^{k} b^j a^{m_j}, \; \sum_{j=1}^{k} b^j a^{m_j} + b^{k+1} a^{m_k}).$$

These intervals are like the intervals in (9) and therefore, by (10), the g-measure of their union (which is evidently a finite union of open intervals) is bounded below by:

$$\sum_i v(r_i) = \sum_{k=1}^{[qn]+1} A \binom{k + m_k - 1}{k-1} q^k p^{m_k}$$

$$(11)\cdots \qquad = Aq \sum_{k=0}^{[qn]} \binom{n}{k} q^k p^{n-k},$$

which tends to $A \cdot q$ as $n \to \infty$. On the other hand, the Lebesgue measure of this union is bounded above by:

$$\sum_{k=1}^{[nq]+1} \binom{n}{k-1} b^{k+1} a^{n+1-k} =$$

$$(12)\cdots \; b^2 \sum_{k=0}^{[nq]} \binom{n}{k} b^k a^{n-k}.$$

Using Stirling's formula, we can show that the last term of this sum is of the order of

$$T_j = \frac{1}{\sqrt{2\Pi pqn}} \left[(\frac{b}{q})^q \; (\frac{a}{p})^p \right]^n$$

and the other terms in this sum can be written as:

$$T_{j-1} = T_j \cdot \frac{[nq]}{n-[nq]+1} \ \frac{a}{b}$$

$$\leq T_j \cdot \frac{nq}{n-nq+1} \cdot \frac{a}{b} \leq T_j \cdot \frac{qa}{pb} = T_j \cdot \lambda$$

and similarly,

$T_{j-2} \leq T_j \cdot \lambda^2$ and so on.

Therefore, the sum (12) is bounded above by $Tj/(1 - \lambda)$, which goes to zero as $n \to \infty$. It follows from (11) that $g(x)$ is <u>not</u> absolutely continuous.

Q.E.D.

We remark that there are values of p, q, a and b such that $0 < pq$, $p + q = 1$, $0 < a \cdot b$, $a + b > 1$ and $(\frac{p}{a})^p \cdot (\frac{q}{b})^q \geq 1$. Notice that if $t(x) = (\frac{p}{x})^p \ (\frac{q}{1-x})^q$, then the minimum of $t(x)$ is attained when $x^p(1-x)^q$ is maximum, i.e. when $x = p$. Hence, we have:

$$t(x) \geq (\frac{p}{p})^p \ (\frac{q}{1-p})^q = 1$$

with equality iff $x = p$. Hence, if $p \neq a$,

$$t(a) = (\frac{p}{a})^p \ (\frac{q}{1-a})^q > 1$$

It is clear that by choosing b slightly greater than $1 - a$, we can have:

$$(\frac{p}{a})^p \ (\frac{q}{b})^q > 1.$$

C. Continuation of The Second Example: The Semigroup of nxn Stochastic Matrices.

Let S be the topological semigroup of nxn stochastic matrices with usual topology and matrix multiplication. Then S is a compact Hausdorff semigroup and its kernel (smallest two-sided ideal) K consists of all stochastic matrices with identical rows, and is a right-zero semigroup. We'll identify K with the set of all probability vectors in R^n. Suppose that μ is a probability measure on S with support B. Let S' be the closed subsemigroup of S generated by B. <u>From this point on, we'll assume throughout that $S' \cap K$ is non-empty</u>. [This condition is easily met if B contains even one stochastic matrix P such that for some positive integer m, the matrix P^m has all its entries positive; for, then by [16,p.139], $\lim P^n$ exists as n tends to infinity and this limit is a matrix in K.] Now the kernel of S' is the set $C=S' \cap K$, which is a right-zero semigroup. By Rosenblatt's theorem [4.13], the sequence of convolution iterates μ^n converge weakly to some probability measure λ whose support is C and which satisfies the convolution equation:

$$\upsilon * \mu = \upsilon. \tag{13}$$

We notice that the equation (13) has a unique solution as before.

Here we consider the same problem as before and wish to determine υ in terms of μ, when μ has a two-point support B. In what follows, $B=\{P,Q\}$, and $\mu(\{P\})=p$ and $\mu(\{Q\})=q$ where $pq > 0$ and $p+q=1$.

Our first result proves a continuity property of υ.

5.11 <u>Proposition</u>. Suppose that both $P^\infty = \lim_{n\to\infty} P^n$ and $Q^\infty = \lim_{n\to\infty} Q^n$ exist and $P^\infty \neq Q^\infty$. Then given $\varepsilon > 0$, there exists $\delta > 0$ such that for any open set V with diameter less than δ, $\upsilon(V) < \varepsilon$.

<u>Proof</u>. Suppose that the proposition is false. Then there exists $\varepsilon > 0$ and a sequence of open spheres V_n such that $\mathrm{diam}(V_n) < 1/n$ and $\upsilon(V_n) > \varepsilon$. Let A be an accumulation point of the sequence of centers of V_n. Then $\upsilon(A) \geq \varepsilon$. Since the mapping $x \to \upsilon(Ax^{-1})$ is upper semicontinuous, it assumes a maximum at some point z in S'. Let $A_1 = Az^{-1}$ and $\upsilon(A_1) = a$. Since υ satisfies equation (13), it is clear that $a \geq \varepsilon$. Since $P^\infty \neq Q^\infty$, either P^∞ or Q^∞ is different from A_1. Suppose $P^\infty \neq A_1$. Now from (13),

$$\upsilon(A_1) = p\upsilon(A_1P^{-1}) + q\upsilon(A_1Q^{-1})$$

and therefore, $\upsilon(A_1P^{-1}) = a$. Similarly, $\upsilon(A_1P^{-n}) = a$ for all positive integers n. Since each one of the sequence A_1P^{-n} is distinct from the other and υ is a probability measure, this gives us a contradiction. The proposition now follows easily.

Our next result is basic to the derivation of Theorem 5.13, the main result in this context.

Let D be the free semigroup with identity I generated by $\{T_1, T_2\}$ where T_1 and T_2 are mappings from the set of probability vectors in R^n into itself such that $xT_1 = x.P$ and $xT_2 = x.Q$. For s in D, let $|s|$ denote the length of s. By $t \nmid s$, we mean that t, s are in D and t is a proper right divisor of s, i.e. $t_1 t = s$ for some $t_1 \neq I$ in D. Then we have the following result.

5.12 Proposition. Suppose that matrices $P=(p_{ij})$ and $Q=(q_{ij})$ satisfy the following condition: for all i, j

$$(14) \qquad \sum_{k=1}^{n} |p_{ik}-p_{jk}| < 2 \text{ and } \sum_{k=1}^{n} |q_{ik}-q_{jk}| < 2.$$

Then there exists a constant r in (0,1) such that for any probability vectors x, y and for any s in D with $|s| = m$, we have:

$$(15) \qquad d(xs,ys) \leq r^m.d(x,y)$$

where for $x=(x_i)$ and $y=(y_i)$, $d(x,y) = \sum_{i=1}^{n} |x_i - y_i|$.

Proof. It is sufficient to prove that for any vector $c=(c_1,c_2,\ldots,c_n)$ in R^n with $\sum_{i=1}^{n} c_i = 0$, we have

$$(16) \qquad \sum_{j=1}^{n} \left(\left|\sum_{i=1}^{n} c_i p_{ij}\right| \right) \leq \frac{\delta}{2} \cdot \sum_{j=1}^{n} |c_{ij}| \quad \text{and}$$

$$(17) \qquad \sum_{j=1}^{n} \left(\left|\sum_{i=1}^{n} c_i q_{ij}\right| \right) \leq \frac{\delta}{2} \cdot \sum_{j=1}^{n} |c_{ij}|,$$

where $\delta = \sup_{i,j} \{ \sum_{k=1}^{n} |p_{ik}-p_{jk}|, \quad \sum_{k=1}^{n} |q_{ik}-q_{jk}| \} < 2.$

We'll prove only (16), since the proof of (17) is similar. If all c_i are zero, then (16) is immediate. Otherwise, we can find real numbers a_{il} such that (i) for each i, $c_i = \sum_l a_{il}$; (ii) for a fixed i, all the a_{il} are of the same sign and (iii) if k many a_{il} are equal to a number b, then there are exactly another k a_{il}'s which are equal to -b. Then we have:

$$\sum_j \left| \sum_i c_i p_{ij} \right| \leq \sum_{a_{il}>0} a_{il} \sum_{j=1}^{n} |p_{ij} - p_{k(i)j}|$$

$$\leq \delta \sum_{a_{il}>0} a_{il} \leq \frac{\delta}{2} \cdot \sum_{i=1}^{n} |c_i|.$$

This completes the proof.

Now we make the following definitions:

(i) if $s \in D$ and s is the product (in any order) of k T_1's and m T_2's, then we define: $v(s)=p^k q^m$;

(ii) for any open set $O \subset R^n$, we define:

(18) $$S(O) = \{s: s \in D, Cs \subset O \text{ and } t \nmid s \text{ implies } Ct \not\subset O\}.$$

Then we have the following theorem.

5.13 Theorem. Suppose that the matrices P and Q satisfy the condition (14). For every open set $O \subset R^n$, let $f(O) = \sum_{s \in S(O)} v(s)$.

Then for all open sets $O \subset R^n$, $f(O) = \lambda(O)$.

Proof. The proof of this theorem will follow from the following lemmas, where we'll show that f can be extended to a regular probability measure with support C and satisfying the convolution equation (13). The theorem will follow from this fact, since the solution of (13) is unique.

LEMMA 1. For every open set $O \subset R^n$, f satisfies:

(18) $$f(O)=pf(OP^{-1})+qf(OQ^{-1}).$$

Proof. It is clear that $C \subset O$ implies that $C \subset OP^{-1}$ and $C \subset OQ^{-1}$, and in this case, $I \in S(O) \cap S(OP^{-1}) \cap S(OQ^{-1})$ and so (18) follows. Otherwise, it can be verified easily that $S(O)=S(OP^{-1})T_1 \cup S(OQ^{-1})T_2$, and that the union on the right-hand side is a disjoint union. From this observation, (18) follows immediately.

LEMMA 2. For every non-negative integer m, $C = \bigcup_{|s|=m} Cs$.

Proof. The lemma is trivial for m=0. We notice that the set $CP \cup CQ$ is contained in C, and also is an ideal of C. Therefore, $CP \cup CQ = C$. Hence

$$\bigcup_{|s|=m+1} Cs = \bigcup_{|s|=m} (CP \cup CQ)s = \bigcup_{|s|=m} Cs.$$

The lemma now follows easily by induction.

LEMMA 3. For every open set $O \subset R^n$ such that $O \cap C$ is non-empty, f(O) is positive.

Proof. Let $x \in O \cap C$. Then there is a positive ε such that $d(y,x) < \varepsilon$ implies $y \in O$. Since for a sufficiently large positive integer m and s in D with $|s| = m$, the diameter of Cs is less than ε by Proposition 5.12, it follows from Lemma 2 that for some s in D, $Cs \subset O$. Hence S(O) is non-empty, and consequently, f(O) is positive.

LEMMA 4. Suppose that O_1 and O_2 are open sets in R^n such that $O_1 \subset O_2$. Then $f(O_1) \leq f(O_2)$.

Proof. It is clear that if $s \in S(O_1)$, then either $s \in S(O_2)$ or there exists $t \nmid s$ such that $t \in S(O_2)$, in which case $v(s) \leq v(t)$. Also, even if the set

$$A_t = \{s \in S(O_1) : t \nmid s\} \text{ for } t \in S(O_2)$$

contains more than one element, then it is easy to verify that

$$\sum_{s \in A_t} v(s) \leq v(t).$$

The lemma now follows from this observation.

LEMMA 5. Suppose that O_1 and O_2 are any two open sets in R^n such that $d(O_1, O_2)$ is positive. Then $f(O_1 \cup O_2) \geq f(O_1) + f(O_2)$.

Proof. Clearly $S(O_1)\cap S(O_2)$ is empty. Now for s in $S(O_1)$, $Cs\subset O_1\subset O_1\cup O_2$. Therefore, either $s\in S(O_1\cup O_2)$ or there exists $t\neq s$ such that $t\in S(O_1\cup O_2)$. Since as in Lemma 4, $\Sigma\{v(s):t\neq s\}\leq v(t)$, the lemma follows.

LEMMA 6. Define for $A\subset R^n$, $f^*(A) = \inf\{f(O):$ O open and $A\subset O\}$. Then (i) f^* is an outer measure; (ii) for open O, $f^*(O)=f(O)$; (iii) f^* is a metric outer measure i.e. $d(A,E)>0$ implies $f^*(A\cup E) = f^*(A)+f^*(E)$ for any two sets A and $E\subset R^n$.

Proof. Assertion (i) follows immediately from the definition of f^*. Assertion (ii) follows from Lemma 4. To prove (iii), let $\varepsilon > 0$. Then there exists an open set $O\supset A\cup E$ such that

$$f^*(O) \leq f^*(A\cup E) + \varepsilon.$$

Since $d(A,E)$ is positive, we can find open sets O_1, O_2 such that $O_1\supset A$, $O_2\supset E$, $O_1\cup O_2\subset O$ and $d(O_1,O_2)$ is positive. Now by Lemmas 4 and 5,

$$f^*(O) \geq f^*(O_1\cup O_2) \geq f(O_1)+f(O_2) \geq f^*(A)+f^*(E).$$

Since $\varepsilon > 0$ is arbitrarily chosen and f^* is an outer measure, (iii) follows.

LEMMA 7. The outer measure f^*, restricted to the Borel subsets of C, is a regular probability measure υ such that $f^*(O)=f(O)$ for all open sets O. Moreover, $\upsilon*\mu=\upsilon$ and υ has support C. Hence, $\upsilon=\lambda$.

Proof. By Lemma 6, f^* is a metric outer measure. By [56,p.59], all Borel subsets of C are f^*-measurable. Hence the restriction of f^* is a regular probability measure on the Borel subsets of C. The rest follows from Lemmas 1 and 3.

Comments on the results of section 5.

The example on $[0,\infty)$ in 5A is taken from A. Mukherjea and E. B. Saff [50]. The example on 2x2 stochastic matrices is first mentioned in M. Rosenblatt [67] & [42]. In [67], questions on absolute continuity of the limit measure are asked. These questions are completely answered in the case $a + b \leq 1$ by T. C. Sun [70]. The case $a + b < 1$ and other related results are taken here from J. R. Gard and A. Mukherjea [17A]. The case $a + b = 1$ as well as the difficult case $a + b > 1$ has been studied by A. Nakassis. Propositions 5.7, 5.8, 5.9 and 5.10 are all his results. The generalization of the second example to the case of nxn stochastic matrices is due to A. Mukherjea and A. Nakassis. All the results in this case are taken from their paper [49].

REFERENCES

1. Argabright, L. N., A note on invariant integrals on locally compact semigroups, Proc. Amer. Math. Soc. 17, 1966, 377-382.

2. Berglund, J. F. and K. H. Hofmann, Compact semitopological semigroups and weakly almost periodic functions, Lecture Notes in Math. no. 42, Berlin-Heidelberg-New York, Springer 1967.

3. Bhattacharya, R. N., Speed of Convergence of the n-fold convolution of a probability measure on a compact group, Z. Wahrscheinbichkeits theorie verw. Geb. 25, 1972, 1-10.

4. Bourbaki, N., Elements de Mathematique: (livre VI, fasc. 29, Chaptitre VII) Hermann, Paris, 1963.

5. Center, B. and A. Mukherjea, More on limit theorems for probability measures on semigroups and groups, Z. Wahrscheinlichkeits theorie verw. Gebiete, to appear.

6. Choy, S. T. L., Idempotent measures on compact semigroups, Proc. London Math. Soc. (3), 30, 1970, 717-733.

7. Choquet, G. and J. Deny, Sur l'equation de convolution C. R. Acad. Sci. Paris, Ser. A-B 250, 1960, 799-801.

8. Clifford, A. H. and G. B. Preston, The algebraic theory of semigroups, Math. surveys No. 7, Amer. Math. Soc., Providence, R.I., 1961.

9. Collins, H. S., Idempotent measures on compact semigroups, Proc. Amer. Math. Soc. 13, 1962, 442-446.

10. Collins, H. S., The Kernel of a semigroup of measures, Duke Math. Journal 28, 1961, 387-92.

11. Collins, H. S., Primitive idempotents in the semigroup of measures, Duke Math. Journal 28, 1961, 397-400.

12. Collins, H. S., Convergence of convolution iterates of measures, Duke Math. Journal 29, 1962, 259-264.

13. Csiszár, I., On infinite products of random elements and infinite convolutions of probability distributions on locally compact groups, Z. Wahrscheinlichkeits - theorie verw. Gebiete 5, 1966, 279-295.

14. Csiszár, I., On the weak* - continuity of convolution in a convolution algebra over an arbitrary topological group, Studia Sci. Math. Hungar. 6, 1971, 27-40.

15. Eckmann, B., Über monothetische gruppen, Comment. Math. Helv. 16, 1943-44, 249-263.

16. Feller, W., An Introduction to Probability theory and its applications, Vol. I, J. Wiley and Sons, 1968.

17. Galmarino, A. R., The equivalence theorem for compositions of independent random elements on locally compact groups and homogeneous spaces, Z. Wahrscheinlichkeits theorie verw. Geb. 7, 1967, 29-42.

17A. Gard, J. R. and A. Mukherjea, On the convolution iterates of a probability measure, Semigroup Forum 10, 1975, 171-184.

18. Gelbaum, B. R. and G. K. Kalisch, Measure in semigroups, Canadian J. Math. 4, 1952, 396-406.

19. Glicksberg, I., Convolution semigroups of measures, Pacific J. Math. 9, 1959, 51-67.

20. Glicksberg, I., Weak compactness and separate continuity, Pacific J. Math. 11, 1961, 205-214.

21. Grenander, U., Probabilities on Algebraic Structures, J. Wiley (New York), 1963.

22. Grosser, S. and M. Moskovitz, On central topological groups, Trans. Amer. Math. Soc. 127, No. 2, 1967, 317-340.

23. Heble, M. and M. Rosenblatt, Idempotent measures on a compact topological semigroups, Proc. Amer. Math. Soc. 14, 1963, 177-184.

24. Hewitt, E. and K. A. Ross, Abstract Harmonic Analysis, Vol. 1 Springer (Berlin, Heidelber, New York), 1963.

25. Hewitt, E. and H. S. Zuckerman, Arithmetic and limit theorems for a class of random variables, Duke Math. Journal 22, 1955, 595-615.

26. Heyer, H., Über Haarche Maße auf lokalkompakten Gruppen, Arch. Math. X VII, 1966, 347-351.

27. Heyer, H., Fourier transforms and probabilities on locally compact groups, Jahresbericht d. DMV 70, 1968, 109-147.

28. Hofmann, K. H. and P. S. Mostert, Elements of compact semigroups, Charles E. Merrill, 1966.

29. Ito, K. and M. Nishio, On the convergence of sums of independent Banach space valued random variables, Osaka Math. J. 5, 1968, 35-48.

30. Kawada, Y. and K. Ito, On the probability distributions on a compact group I, Proc. Phys.-Math. Soc. Japan ser. 3, 22, 1940, 977-998.

31. Kelley, J. L., Averaging operators on C_{α} (X), Illinois M. Math. 2, 1958, 214-223.

32. Koch, R. J., Topological semigroups, Tulane University dissertation, 1953.

33. Kloss, B. M., Probability distributions on bicompact topological groups, Theory of Prob. Applns. 4, 1959, 234-270.

34. Kloss, B. M., Limiting distributions on bicompact abelian groups, Theory of Prob. Applns. 6, 1961, 361-389.

35. de Leeuw, K. and I. Glicksberg, Applications of almost periodic compactifications, Acta Math. 105, 1961, 361-389.

36. Levy, P., Theorie de L'addition des variables aleatoires, Gauthier-Villars (Paris), 1937.

37. Levy, P., L'addition des variables aleatoires definis sur une circomference, Bull. soc. Math France 67, 1939, 1-41.

38. Ljapin, E. S., Semigroups, Amer. Math. Loc. (Translations), 1968.

39. Loeve, M., Probability theory, Van Nostrand (Princeton), 1963.

40. Loynes, R., Probability distributions on a topological group, Z. Wahrscheinlichkeits theorie verw. Gabiete 5, 1966, 446-455.

41. Martin-Löf, P., Probability theory on discrete semigroups, Z. Wahrscheinlichkeits theorie verw. Geb. 4, 1965, 78-102.

42. Maximov, V. M., A Generalized Bernoulli scheme and its limit distributions, Theory of Prob. and its applications, X 18 No. 3, 1973, 521-530.

43. Maximov, V. M., Composition convergent sequences of measures on compact groups, Theory of Prob. and its applications, 16, No. 1, 1971, 55-73.

44. Maximov, V. M., Necessary and sufficient conditions for the convergence of non-identical distributions on a finite group, Theory of Probability and its applications 13, 1968, 287-298.

45. Mukherjea, A., On the convolution equation $P = P * Q$ of Choquet and Deny for probability measures on semi-groups, Proc. Amer. Math. Soc. 32, 1972, 457-463.

46. Mukherjea, A., On the equations $P(B) = \int P(Bx^{-1})P(dx)$ for infinite P, J. London Math. Soc. (2), 6, 1973, 224-230.

47. Mukherjea, A., Limit theorems for probability measures on non-compact groups and semigroups, Z. Wahrscheislichkeits theorie verw. Gebiete 33, 1976, 273-284.

48. Mukherjea, A., Limit theorems for probability measures on compact or completely simple semigroups, Trans. Amer. Math. Soc., to appear.

49. Mukherjea, A. and A. Nakassis, Limit behavior of convolution iterates of probability measures on n x n stochastic matrices, J. Math. Analysis and Applications, to appear.

50. Mukherjea, A. and E. B. Saff, Behavior of convolution sequences of a family of probability measures on $[0,\infty)$, Indiana University Math. J. 24, No. 3, 1974, 221-226.

51. Mukherjea, A. and Tze-chien Sun, Convergence of products of independent random variables with values in a semigroup, to appear.

52. Mukherjea, A. and N. A. Tserpes, Idempotent measures on locally compact semigroups, Proc. Amer. Math. Soc. 29, No. 1 1971, 143-150.

53. Mukherjea, A. and N. A. Tserpes, Invariant measures and the converses of Haar's theorem on semi-topological semigroups, Pacific J. Math. 21, No. 10, 1972, 973-977.

54. Mukherjea, A. and N. A. Tserpes, A problem on r* - invariant measures on locally compact semigroups, Indiana U. Math J. 21, No. 10, 1972, 973-977.

55. Mukherjea, A. and N. A. Tserpes, On certain conjectures on invariant measures on semigroups, Semigroup Forum 1, 1970, 260-266.

56. Munroe, M. E., Measure and Integration (2nd Edition), Addison-Wesley Pub. Co., Reading, Mass., 1971.

57. Numakura, K., On bicompact semigroups, Math. J. Okayana Univ. 1, 1952, 99-108.

58. Parthasarathy, K. R., Probability measures on metric spaces, Academic Press, 1967.

59. Parthasarathy, K. R., R. Ranga Rao and S. R. S. Varadhan, Probability distributions on locally compact abelian groups, Illinois J. Math. 7, 1963, 377-369.

60. Pym, J. S., Idempotent measures on Semigroups, Pacific J. Math. 12, 1962, 685-698.

61. Pym, J. S., Idempotent measures on compact semitopological semigroups, Proc. Amer. Math. Soc. 21, 1969, 499-501.

62. Pym, J. S., Dual structures for measure algebras, Proc. London Math. Soc. (3), 19, 1969, 625-660.

63. Rosen, W. G., On invariant measures over compact semigroups, Proc. Amer. Math. Soc. 7, 1956, 1076-1082.

64. Rosenblatt, M., Limits of convolution sequences of measures on a compact topological semigroup, J. Math. Mech. 9, 1960, 293-306.

65. Rosenblatt, M., Products of independent, Identically distributed stochastic matrices, J. Math. Analysis and Applications, 11, 1965, 1-10.

66. Rosenblatt, M., Equicontinuous Markov operators Theory of Prob. Applns. 9, 1964, 180-197.

67. Rosenblatt, M., Markov Processes: Structure and Asymptotic behavior, Springer (Berlin, Heidelberg, New York), 1971.

68. Sazonov, V. V. and V. N. Tutubalin, Probability distributions on topological groups, Theory of Prob. and Applns. 11, 1966, 1-47.

69. Schwartz, S., Convolution semigroup of measures on compact non-commutative semigroups, Czech. Math. J. 14, (89), 1964, 95-115.

70. Sun, Tze-chien, On the limit of convolution iterates of a probability measure on 2 x 2 matrices, Bulletin of the Institute of Mathematics, Academia Sinica, 1975.

71. Sun, Tze-chien and N. A. Tserpes, Idempotent probability measures on locally compact abelian semigroups, J. Math. Mech. 19, 1970, 1113-1116.

72. Stromberg, K., Probabilities on a compact group, Trans. Amer. Math. Soc. 94, 1960, 295-309.

73. Tserpes, N. A. and A. Kartsatos, Measure semi-invariants sur un semi-groups localement compact, C.R. Acad. Sci. Paris. Ser. A, 267, 1968, 507-509.

74. Tortrat, A., Lois tendues β sur un demi-groups topologique completement simple, Z. Wahrscheinlichkeits theorie verw. Gebiche 6, 1966, 145-160.

75. Tortrat, A., Lois de probabilite sur un espace topologique completement regulier et produits infinis a termes independants dans un groupe topologique, Ann. Inst. Henri Poincare 1, 1965, 217-237.

76. Urbanik, K., On the limiting probability distributions on a compact topological group, Fund. Math. 3, 1957, 253-261.

77. Vorobev, N. N., The addition of independent random variables on finite groups, Mat. Sbornik 34, 1954, 89-126.

78. Wendel, J. G., Haar measures and the semigroups of measures on a compact group, Proc. Amer. Math. Soc. 5, 1954, 923-929.

79. Williamson, J. H., Harmonic analysis on semigroups, J. London Math. Soc. 42, 1967, 1-41.

CHAPTER II

RECURRENT RANDOM WALKS ON TOPOLOGICAL GROUPS AND SEMIGROUPS

1. Introduction

Intimately connected with the convergence behavior of convolution sequences of probability measures μ on locally compact groups and semigroups, is the recurrence and conservativeness concepts for the states of the stationary random walks induced by the initial measure μ and transition functions $P(x,\cdot) = \delta_x * \mu$ (right random walk), $\mu * \delta_x$ (left) and $\mu * \delta_x * \mu$ (bilateral), δ_x being the "point mass" at x. Also, recurrence concepts are connected with the existence of invariant measures for the random walks, i.e., σ-finite measures π satisfying $\pi P = \pi$. The recurrent random walks can be classified into two great types: recurrent null (invariant measure unbounded) and recurrent positive (invariant measure bounded). Our approach in studying recurrence is based on the stationary and independent-increment structure of these walks rather than the potential and functional analytic methods.

The(right) random walk $S_n = X_1 X_2 \cdots X_n$ on a topological group G is said to be recurrent if there is a recurrent value $x \in G$, i.e., a value x satisfying

(1) $\quad P(S_n \in N(x) \text{ i.o.(infinitely often)}) = 1$

for every neighborhood $N(x)$ of x.

(Here the X_i are independent random variables with values in G with common probability law μ). Actually, for recurrence of S_n, it is enough that (1) holds for some compact neighborhood of the identity. A value $x \in G$ is a possible value of the random walk if for every $N(x)$ there is an n such that $P(S_n \varepsilon N(x)) > 0$. When G is the Euclidean k-space, Chung and Fuchs [3] showed: either there are no recurrent values or all possible values are recurrent (and form a closed subgroup). Moreover, S_n is recurrent if for some compact neighborhood V of the identity

$$(2) \qquad \int_V \frac{du}{1 - \phi(u)} = \infty$$

where ϕ is the characteristic function of the distribution of μ. Kesten and Spitzer [10] gave a necessary and sufficient criterion similar to (2) using the character group on a countable locally compact abelian G. This criterion of Kesten and Spitzer was subsequently extended by Port and Stone [22] to the case of a general locally compact abelian group. Besides this nice computational criterion of recurrence, we also have criteria due to Dudley ([6] and [7]) that tells us which abelian groups can support a recurrent random walk. Such criteria for non-abelian countable groups were also sought by Kesten in [11]. Kesten [12] also studied symmetric random walks (i.e., when μ is inversion-invariant) on countable abelian groups and proved that the distinction $\lambda = 1$ or $\lambda > 1$ (where λ is the largest element of the spectrum of the transition matrix) depends only on the group and not on the particular symmetric μ used to define the random walk. (This symmetric case was generalized to completely simple semigroups by Larisse [15]). For non-abelian locally compact groups, Loynes [16] generalized the main results of Chung and Fuchs mentioned above.

The noteworthy divergence between the theory of recurrence on groups and that on semigroups is shown by the fact that (1) is not equivalent to

(3) $P_x(S_n \varepsilon N(x) \text{ i.o.}) = 1$, for all neighborhoods $N(x)$ of x

For example, if $S = G \times Y$, where G is the group $\{1,a\}$, $Y = \{e_1, e_2\}$ is a right-zero semigroup, and each singleton in S has μ-measure $\frac{1}{4}$, then $P(S_n = (1,e_1) \text{ i.o.}) = \frac{1}{2}$ while $P_{(1,e_1)}(S_n = (1,e_1) \text{ i.o.}) = 1$. It turns out that for semigroups it is more natural to call a state recurrent if (3) holds. Henceforth "recurrent" will be interpreted in the sense of (3). Random walks on compact (resp. countable locally compact) semigroups were considered by Rosenblatt [25] (resp. [26]), in his formulation of necessary and sufficient conditions for the convergence of the unaveraged (and averaged) convolution sequence μ^n. Rosenblatt's book [24] contains a wealth of information and motivation for random walks on semigroups. Also the book of Grenander [8] contains some related material mainly on groups. Martin-Löf [17] considered the same problem as Rosenblatt using the unilateral (left and right) walks on a discrete (countable) semigroup S. Using the terminology of Rosenblatt [28] let us call a state x conservative if

(4) $\sum P^n(x,N(x)) = \infty$, for all neighborhoods $N(x)$ of x
(for the discrete case read $\{x\}$ for $N(x)$)

In the discrete (countable) case (where Markov chain techniques are available), a state x is conservative if and only if it is recurrent (in the sense of (3)). Martin-Löf proved: If one walk is conservative (= recurrent), so is the other and the recurrent states are either all null or all positive and they constitute the completely simple minimal ideal (= kernel) of S. The results obtained by Martin-Löf were completed by Larisse [14] (still working in the discrete case) who introduced also the bilateral walk and proved that recurrence in the bilateral is equivalent to recurrence in the unilateral walks. He also proved that the essential classes for the bilateral walk are at most two in the recurrent case.

Contrary to the case of a countable Markov chain, a conservative state may not be recurrent as Rosenblatt has recently given an example in [27]. In general the methods in the "continuous" (non-discrete) case are markedly different from those (based on Markov chain tools) of the discrete case. In a series of papers of the authors and T.C. Sun, recurrent random walks in the continuous case were considered for completely simple semigroups, compact semigroups and locally compact abelian groups ([29], [18], [19] and [20] respectively). The main results were: (i) The equivalence of "conservativeness" and recurrence. (ii) The recurrent states form the completely simple minimal ideal of S and in the compact case all the three random walks are recurrent. (iii) The equivalence between unilateral and bilateral recurrence for compact semigroups and locally compact abelian groups. (iv) A completely simple semigroup $S = E \times G \times F$ can support a recurrent (left) walk if and only if the group factor G can support such a walk.

Recently, bilateral and more general walks on compact semigroups were considered independently by Högnäs [9] who utilized techniques in Rosenblatt's book [24] to prove necessary and sufficient conditions for the random walks to be ergodic and mixing. He also determined (independently) the structure of essential (=recurrent) classes for the bilateral walk. Also, Brunel and Revuz [2], using a theorem of Brunel giving new conditions for the existence of a bounded invariant measure, proved: If every random walk on a locally compact metrizable group is recurrent, then the group is compact.

2. Notations

2.1 Preliminaries: Throughout this chapter S will be (at least) a locally compact Hausdorff 2nd countable topological semigroup and μ a regular Borel probability measure with support $S(\mu) \equiv S_\mu$. For $A, B \subset S$, $x \in S$, $\bar{A} \equiv cl(A) =$ closure (A), A^c = complement of A, 1_A = the indicator function of A, and

$$AB^{-1} = \{s \in S;\ \text{there is } b \in B \text{ such that } sb \in A\}$$

$Ax^{-1} = \{s \in S;\ sx \in A\}$, similar definitions holding for $A^{-1}B$ and $x^{-1}A$. Let

$D = \overline{\bigcup_{n=1}^{\infty} S_\mu^n}$, and $\mathcal{B}$ the Borel σ-field of D generated by the open sets.

The set D contains all possible states of the three random walks (left, right, bilateral) induced by μ having initial measure μ and transition functions $P(x,B)$ given resp. by

$$P(x,B) = \begin{cases} P_r(x,B) = \mu(x^{-1}B) \equiv {}_x\mu(B) = \delta_x * \mu\ (B) \\ P_\ell(x,B) = \mu(Bx^{-1}) \equiv \mu_x(B) = \mu * \delta_x(B) \\ P_b(x,B) = [\mu * \mu(x^{-1}\cdot)]\ (B) \equiv \mu *_x \mu\ (B) = \mu * \delta_x * \mu \end{cases}$$

the "iterates" of these transition functions being

$$P^n(x,B) = \left| \begin{array}{l} P_r^n(x,B) = \mu^n(x^{-1}B) = \delta_x * \mu^n\ (B) \\ \\ P_b^n(x,B) = [\mu^n * \mu^n(x^{-1}\cdot)]\ (B) = \mu^n * \delta_x * \mu^n\ (B) \end{array} \right.$$

where $*$ denotes convolution, $\mu^n = \mu * \mu * .. * \mu$ (n times) and δ_x is the "point-mass" at $\{x\}$.

For the significance of the bilateral walk, introduced in [14] we observe that when D is discrete, the transition matrix for the bilateral walk is the (commutative) product of the transition matrices for the left and the right walk. In the continuous case we have

$$P_b(x,B) = \int P_r(z,B)P_\ell(x,dz) = \int P_\ell(z,B)\, P_r(x,dz)$$

2.2 Representation of the random walks:

It turns out that the left, right and bilateral random walks can be represented as products

$$S_n = \begin{cases} Z_n = X_1X_2\cdots X_n & \text{(right)} \\ L_n = X_nX_{n-1}\cdots X_1 & \text{(left)} \\ W_n = X_{-n}\cdots X_{-1}X_oX_1\cdots X_n & \text{(bilateral)} \end{cases}$$

where X_o, $X_{\pm 1}$, $X_{\pm 2}$ .., are independent random variables with values in S identically distributed according to μ. More generally X_o (or X_1) may have an initial measure $\mu_o \neq \mu$ whose support is contained in D. For our treatment however, we shall assume $\mu_o = \mu$.

The X_i's can be realized as the usual coordinate functions $X_i(\omega) = d_i$ on sequences $\omega = (d_1, d_2, ..)$ of the product measure space $(D^\infty \equiv \Pi\, D_i, P)$, $D_i = D$ for all i, with the product measure P induced by μ.

P_x will denote the product measure induced by μ and the initial measure δ_x.

Our "blanket" assumption of 2nd countability makes the S_n automatically measurable. In almost all of our results we can remove 2nd-countability by taking (D^∞, P) to be the Bledsoe-Morse product measure extension of the usual product measure space, as utilized by Dudley [7, p.233-234]. Then the S_n's would be measurable (Borel) as functions on (D^∞, P).

2.3 Whenever it is not necessary to specify which random walk is considered, we will use the generic S_n (S_o= X_1 or X_o as the case requires) to indicate any of Z_n, L_n, W_n and $P^n(x,\cdot)$ to indicate any of the three transition functions.

A point $x \varepsilon S$ is called "possible" for one of the random walks if for each (open) neighborhood $N(x) \equiv N_x$ of x, there is $n \geq 0$ such that $P(S_n \varepsilon N_x) > 0$.

We observe that D is exactly the set of possible points (= states) for each of the unilateral (left and right) random walks. (D contains the possible states for W_n).

2.4 Communication Relations: We say that $x \varepsilon S$ leads to $y \varepsilon S$ in the left random walk ($x \to y$) if $y \varepsilon \overline{xD}$, and similarly for the right random walk using $\overline{Dx}$. For $x,y \varepsilon S$, we say that $x \to y$ in the bilateral random walk if $y \varepsilon \overline{\bigcup_{n=1}^{\infty} S_\mu^n x S_\mu^n}$. The state x is called essential if $x \to y$ implies $y \to x$.

2.5 Recurrence: For a given random walk S_n and for x, $y \varepsilon D$ we write $x \to y$ i.o. (infinitely often) or $x \to y$ i.o. (S_n) if

(i) $P(S_n \varepsilon N(y) \text{ i.o.} | S_o = x) = P_x(S_n \varepsilon N(y) \text{ i.o.}) = 1$

for every neighborhood N_y of y.

If (i) holds, we sometimes say that y is x-recurrent.

We call $x \varepsilon D$ recurrent if $x \to x$ i.o.

Following Rosenblatt [28], we call $x \varepsilon D$ conservative if

(ii) $\sum P^n(x,N(x)) = \infty$, for all neighborhoods N_x of x.

A point x is called unconditionally recurrent if

(iii) $P(S_n \varepsilon N(x) \text{ i.o.}) = 1$ for all $N(x)$ of x.

Clearly such a point must be in D. Let R^{ℓ}, R^{r}, R^{b} (resp. $R_u^{\ell}, R_u^{r}, R_u^{b}$) be the sets of recurrent states (resp. unconditionally recurrent) for the left, right and bilateral walk, respectively.

A state $x \in D$ is called <u>recurrent positive</u> (resp. <u>recurrent null</u>) if in addition to $x \to x$ i.o., we have $\overline{\lim_n} P^n(x,N(x)) > 0$ for every neighborhood $N(x)$ of x (resp. $\lim_n P^n(x,N(x)) = 0$ for a neighborhood $N(x)$).

A point $x \in D$ is called a <u>point of sure return</u> if

(iv) $P_x(S_n \in N(x)$ for some $n) = 1$, for all $N(x)$ of x.

2.6 <u>Remark:</u> The function $x \to P(x,U)$ (for all the three walks), where U is a fixed open set in D, can be proven to be lower semicontinuous following similar argument as in Rosenblatt [24, p. 130].

$P(x,\cdot)$ defines an operator on continuous functions by

$$n \geq 1, \quad P^n f = \int f(s)\, P^n(x,ds) = \begin{cases} \int f(xs)\mu^n(ds) \\ \int f(sx)\,\mu^n(ds) \\ \iint f(sxt)\mu^n(ds)\mu^n(dt) \end{cases}$$

according as $P \equiv P^1$ is the left, right or bilateral transition function. Since μ is regular, P maps continuous bounded functions into continuous bounded functions.

If S satisfies the conditions that AB^{-1} and $A^{-1}B$ are compact whenever $A, B \subset D$ are compact, then for $f \in C_\infty$ (= the space of continuous functions vanishing at ∞), we have $Pf \in C_\infty$. The above mentioned compactness conditions imply the following (weaker) conditions

(CR) $x \notin Sy$ implies there exist neighborhoods N_x, N_y such that $N_x N_y^{-1} = \emptyset$

(CL) $x \notin yS$ implies there exist N_x, N_y such that $N_y^{-1}N_x = \emptyset$.

These conditions were first introduced by T. C. Sun to obtain certain results on recurrence of random walks for general semigroups. Later these conditions were used in [21] to prove that the collection of points x for which $\Sigma\, \mu^n(N(x)) = \infty$ for every open neighborhood of x, is the completely simple kernel of D.

3. The unilateral walks in the case of completely simple semigroups.

In this section we study recurrence on completely simple semigroups. Such a semigroup has the product-topology structure $S = E \times G \times F$ where G is a group. (See Chapter 1 where the pertinent facts are given). The following theorem follows from the results of Loynes [16]. It generalizes a corresponding result of Chung and Fuchs on Euclidean d-space [3] and rounds up the situation concerning recurrence of the unilateral walks in the group case. Its generalization to completely simple semigroups will be given in this section.

3.1 Theorem. Let G be a group. Then, either no value in D is recurrent in any of the unilateral walks or the following equivalent statements hold:

(a) $R_u^r = R_u^\ell = R^r = R^\ell = D$ = a (closed) subgroup of G

(b) $\Sigma_{n=1}^{\infty} P(Z_n \in N) = \infty$ for all neighborhoods N of the identity e.

(c) $\Sigma_{n=1}^{\infty} P_e(Z_n \in N) = \sum_{n=1}^{\infty} P(Z_n \in N) = \infty$ for some compact neighborhood N of e.

(d) $\sum_{n=1}^{\infty} P_x(Z_n \in N) = \infty$ for all x and open sets $N \neq \phi$.

(e) $P_x(Z_n \in N \text{ i.o.}) = 1$ for all open sets $N \neq \phi$.

3.2 Note: It will be seen in this section that in the recurrent case equation (a) loses members when we go from groups to right groups in which case $R_u^r = R^r = R^\ell = D$ but R_u^ℓ may be empty. Going even further to completely simple semigroups, equation (a) becomes $R^r = R^\ell = D$.

Proof of Theorem: We shall use an argument due to Chung [4 , p. 266]. Recall from the definition (2.1) that D, the set of all possible states for the unilateral walks, is a (closed) subsemigroup. Suppose $R_u^r \neq \phi$: To prove that $R_u^r = D$ = a group, it suffices to show that if x is a possible state and $y \in R_u^r$, then $x^{-1}y \in R_u^r$. Suppose not; then there must be an m and a neighborhood $N(e) = N$ of the identity such that

(i) $P(Z_n \notin x^{-1}yN \text{ for all } n \geq m) > 0.$

Let $N_1(e) = N_1$ be such that $x^{-1}yN \supset N_1^{-1}x^{-1}yN_1$. Since x is possible there is k such that $P(Z_k \in xN_1) > 0$. Now,

$$\{Z_k \in xN_1\} \cap \{Z_k^{-1}Z_n \notin x^{-1}yN \text{ for all } n \geq k+m\} \subset \{Z_n \notin yN_1 \text{ for all } n \geq k + m\}$$

Hence,

(ii) $P(Z_n \notin yN_1 \text{ for all } n \geq k + m)$

$$\geq P(Z_k \in xN_1) \cdot P(Z_k^{-1} Z_n \notin x^{-1}yN \text{ for all } n \geq k + m)$$

Since $Z_k^{-1}Z_n$ and Z_{n-k} have the same probability law, the last probability on the right of (ii) equals the probability in (i). It follows that the first term in (ii) is positive, contradicting the assumption that $y \in R_u^r$. We have thus proved that $R_u^r = D$ is a subgroup of G. (Note that $e = y^{-1}y \in R_u^r$ by the above argument). By the Borel-Cantelli lemma, (a)$\Longrightarrow$(b).

From the above proof, the central role to recurrence of the identity e is apparent. The remaining part of the equation (a) will follow as soon as part (b) is proven together with the fact that

(iii) $P_x(Z_n \in N \text{ i.o.}) = P(xZ_n \in N \text{ i.o.})$

(b) ⟹ (a): Suppose that (b) holds and write

$$q(N) = P(Z_n \notin N \text{ for all } n)$$

$$r(N) = P(Z_n \in N \text{ f.o. } (= \text{for only finitely many } n\ ))$$

for any neighborhood N of e.

Let N be an arbitrary neighborhood and N_n a sequence of symmetric neighborhoods with $N_{n+1}^2 \subset N_n$, $N_1^2 \subset N$. Then,

$$1 \geq r(N_1) \geq \sum_k P(Z_k \in N_1,\ Z_k^{-1} Z_{k+n} \notin N \text{ for all } n)$$

$$\geq \sum_k P(Z_k \in N_1) \cdot q(N)$$

because $Z_k^{-1} Z_{k+n}$ is independent of Z_k and has the same distribution as Z_n. Hence $q(N) = 0$ for any N.

Now write $M_n = N_1 N_2 N_3 \ldots N_n$ and $M = \cup M_n$.

Then each M_n is a neighborhood contained in N, and $M_n \cdot N_{n+1} = M_{n+1}$. We have,

$$r(M) = P(\bigcup_k (Z_k \in M,\ Z_{k+n} \notin M,\ n \geq 1)$$

$$= \lim_j \sum_k P(Z_k \in M_j,\ Z_{k+n} \notin M,\ n \geq 1)$$

$$\leq \lim_j \sum_k P(Z_k \in M_j) \cdot P(Z_k^{-1} Z_{k+n} \notin N_{j+1},\ n \geq 1)$$

$$= 0.$$

Hence, $r(N)$ is zero for any N, and the identity e is recurrent in R_u^r. Using dual argument for the left random walk we can show that $e \in R_u^\ell$.

(c) ⟹ (a). A simple compactness argument shows that there must be $y \in N$ such that

$$\sum_n P_x(Z_n \in N_y) = \infty \quad \text{for all } N_y.$$

or

(iv) $$\sum_n P(xZ_n \in N_y) = \sum_n P(Z_n \in x^{-1} N_y) = \infty$$

which implies that $x^{-1} y \in R_u^r$.

In Euclidean d-space (c) ==> (b) follows also from an interesting lemma proven in [4, p. 268 , Lemma 1], which states: For any $\varepsilon > 0$ and positive integer m, let $N_e(m\varepsilon)$ be the neighborhood of e (with compact closure) consisting of the points at a distance from e less than $m\varepsilon$. Then,

$$\sum_n P(Z_n \in N_e(m\varepsilon)) \leq 2m \sum_n P(Z_n \in N_e(\varepsilon)).$$

3.3 Lemma: Let $S = E\times G\times F$ be completely simple. If $R_u^r \neq \phi$, then D is contained in a sub-right-group of S.

Proof: We observe that if $S = E\times G\times F$ and $x = (e,g,f)\in S$, then $xS = \{e\} \times G \times F$ is a right group.

Suppose $(e,g,f)\in R_u^r$ and $(e',g',f') \in D$ with $e \neq e'$. Then there exist neighborhoods N_e, $N_{e'}$ of e and e' in E respectively such that $N_e \cap N_{e'} = \phi$. Let N_g, $N_{g'}$ be neighborhoods of g and g' in G and N_f and $N_{f'}$ be neighborhoods in F. Since $(e', g', f') \in D$, we have

$\mu^k(N_{e'} \times N_{g'} \times N_{f'}) > 0$ for some $k \geq 1$. Also,

$$N_e \cap N_{e'} = \phi \Rightarrow (N_{e'}\times N_{g'}\times N_{f'})^{-1} (N_e \times N_g \times N_f) = \phi.$$

It follows that

$P(Z_n \in N_e \times N_g \times N_f$ f.o. (finitely often))

$\geq \mu^k (N_{e'} \times N_{g'} \times N_{f'}) > 0.$

Therefore, $(e,g,f)\notin R_u^r$, a contradiction. Hence, there does not exist in D a state (e', g', f') with $e \neq e'$, i.e., $D\subset\{e\} \times G \times F$, a right group. Q.E.D.

3.4 Theorem: If S is completely simple, then either $R_u^r = \phi$ or $R_u^r = D =$ a topological right group in S.

Proof: Since D is inside a right group (cf. 3.3) and since a right group with multiplication from the right behaves like a group, the proof given in 3.1 can be modified to apply in this case. In fact one may assume that D itself is a right group by 3.3; if $R_u^r \neq \phi$ then every $d \in D$ has the property that $\sum_n \mu^n(N_d) = \infty$ for all N_d since the set of points with this property is an ideal and D is simple; then using similar argument as in 3.1, it follows that if $(g,e.) \in R_u^r$ and $(c,e) \in D$, then $(c^{-1}g, e.) \in R_u^r$ and R_u^r is a left ideal. Also by the argument in (3.1, (b)$\Rightarrow$(a)), every idempotent $e \in D$ is in R_u^r, from which one obtains $R_u^r = D$. Q.E.D.

Following the arguments in [3] and those in the proof of 3.1 above, we can also show the following.

3.5 Theorem: If S is a right group, then $R_u^r \neq \phi$ if and only if $\sum_{n=1}^{\infty} P(Z_n \in N_a) = \sum \mu^n(N_a) = \infty$ for all neighborhoods N_a of some $a \in D$.

As it was pointed out in Section 1, it is more reasonable to call, for random walks on semigroups, a state x recurrent if $x \in R^r \equiv \{x \in D: P_x(Z_n \in N_x \text{ i.o.}) = 1$ for all neighborhoods N_x of $x\}$.

3.6 Definition: (i) We say that the right random walk is recurrent if $R^r \neq \phi$. (ii) In the sequel we shall use the notation $Z_k^{-1}Z_n \equiv X_{k+1} \cdot X_{k+2} \cdots X_n$.

3.7 Lemma: In any semigroup S, whenever $R^r \neq \phi$, R^r is a left ideal of D.

It follows from the fact that $P_x(Z_n \in N_x \text{ i.o.}) = 1 \Rightarrow$ $P_x(xZ_n \in N_x \text{ i.o.}) = 1$ which implies $P(yxZ_n \in yN_x \text{ i.o.}) = 1$ and every neighborhood $N_{yx} \supset yN_x$ for some N_x.

In the remaining of this section S will be completely simple, $S = E \times G \times F$. We shall need the following lemmas of which the first one is purely algebraic.

3.8 Lemma: If a subsemigroup D of a completely simple semigroup $S = E \times G \times F$ has a minimal right (or left) ideal (of itself), then D is also completely simple.

Proof: By hypothesis, for some $a \in D$, aD is right simple and being left cancellative ($aD \subset aS$ = a right group), aD is a right group and contains an idempotent. By [5, II, p. 88 or I, p. 84], D has a completely simple kernel $K = E' \times G' \times F'$, $E' \subset E$, $F' \subset F$, $G' \subset G$, and G' is a group. Let $(e,g,f) \in D$. Then $e \in E'$, $f \in F'$ by the ideal property of K. Hence, $fe \in G'$ and so $(fe)^{-1} \in G'$ so that $(e, (fe)^{-1}, f) \in K$ and also $(e,g,f)\cdot(e,(fe)^{-1}, f) = (e,g,f) \in K$. Hence, $K = D$.

3.9 Lemma: Let E be compact. Then for each (neighborhood) N_u of the unit element u in G and for each $f \in F$ there exists N_f of f in F such that $N_f e \subset N_u(fe)$ for all $e \in E$.

Proof: Let $N \equiv N_u$ and $f \in F$ be given. For each $e \in E$, there exists N_{fe} of fe in G, N_f in F, N_e in E, such that:

$N_{fe} N_{fe}^{-1} \subset N$, $N_f N_e \subset N_{fe}$ and hence, for every $e' \in N_e$,

$fe' \subset N_{fe}$, $N_{fe} (fe')^{-1} \subset N$, $N_f N_e \subset N_{fe} \subset N(fe')$ for

all $e' \in N_e$.

Since a finite collection of N_e's cover E, say $N_{e1}, N_{e2}, \ldots N_{e_n}$, then the desired neighborhood is $N_f = \bigcap N_f^{(i)}$, where $N_f^{(i)}$ is the neighborhood of f in F chosen above relative to N_{e_i}, $i = 1, 2, \ldots n$. Q.E.D.

We recall from Section 2 that $x \to y$ i.o. means that $P_x(Z_n \in N_y$ i.o.$) = 1$ for every (neighborhood) N_y of y. Also $x \to y$ means $y \in \overline{xD}$.

In the sequel we shall use the notation $Z_k^{-1} Z_n \equiv X_{k+1}X_{k+2}\ldots X_n$.

3.10 Proposition: If $x \longrightarrow y$ i.o. and $x \longrightarrow z$, then $z \longrightarrow y$ i.o.

Proof: (i) Suppose $z \not\rightarrow y$ i.o. Then $P_z(Z_n \in \tilde{N}_y$ f.o.$) =$ $P(Z_n \in z^{-1}\tilde{N}_y$ f.o.$) > \epsilon > 0$ for some neighborhood $\tilde{N}_y$ of y. Choose $H = C \times G \times F$ where $C \subset E$ is compact such that $P(Z_1 \in H^c) < \epsilon/2$.

This implies that $P(Z_n \in H^c \text{ i.o.}) < \varepsilon/2$, since the set $\{Z_n \in H^c \text{ i.o.}\}$ is contained in $\{Z_1 \in H^c\}$.

(ii) Suppose $x = (e,g_x,f_x)$, $y = (e,g_y,f_y)$, $z = (e,g_z,f_z)$, and $\tilde{N}_y = N_e \times \tilde{N}_{g_y} \times \tilde{N}_{f_y}$, $N_y = N_e \times N_{g_y} \times N_{f_y}$, $N_z = N_e \times N_{g_z} \times N_{f_z}$, where N_e, N_{f_y}, $\tilde{N}_{f_y}$ and N_{f_z} are neighborhoods of e, f_y and f_z in E and F respectively, and $N_{g_y} = gyN_u$, $\tilde{N}_{g_y} = g_y\tilde{N}_u$ and $N_{g_z} = g_z\hat{N}_u$ with N_u, $\tilde{N}_u$, $\hat{N}_u$ being neighborhoods of the identity u of G. It is easy to compute

(1) $z^{-1}\tilde{N}_y = \{(a,b,c) \in S;\ c \in \tilde{N}_{f_y} \text{ and } b \in (f_z a)^{-1} g_z^{-1} g_y \tilde{N}_u\}$,

(2) $N_z^{-1}N_y = \{(a,b,c) \in S;\ c \in N_{f_y} \text{ and } b \in (N_{f_z} a)^{-1}\hat{N}_u^{-1} g_z^{-1} g_y N_u$.

(iii) Write $(z^{-1}\tilde{N}_y) \cap H = A$, $(N_z^{-1}N_y) \cap H = B$. Since C is compact, by (1) and (w) and Lemma (3.9), for given z and $\tilde{N}_y$, we can find N_{f_y}, N_{f_z}, $\hat{N}_u$ and N_u such that $B \subset A$.

(iv) Since $x \not\to z$, there exists $k > 0$ such that $P_x(Z_k \in N_z) = P(xZ_{k-1} \in N_z) > 0$. Then

$$
\begin{aligned}
P_x(Z_n \in N_y \text{ f.o.}) &= P(xZ_n \in N_y \text{ f.o.}) \\
&\geq P(xZ_{k-1} \in N_z \text{ and } (xZ_{k-1})^{-1}(xZ_n) \in (xZ_{k-1})^{-1}N_y \text{ f.o.}) \\
&\geq P(xZ_{k-1} \in N_z \text{ and } Z_k^{-1}Z_n \in N_z^{-1}N_y \text{ f.o.}) \\
&\geq P(xZ_{k-1} \in N_z) \cdot P(Z_n \in N^{-1}N_y \text{ f.o.}) \\
&\geq P\ (xZ_{k-1} \in N_z)\ [P(Z_n \in N_z^{-1}N_y \cap H \text{ f.o.}) - P(Z_n \in H^c \text{i.o.})] \\
&\geq P(xZ_{k-1} \in N_z)\ [P(Z_n \in B \text{ f.o.}) - \varepsilon/2] \\
&\geq P(xZ_{k-1} \in N_z)\ [P(Z_n \in A \text{ f.o.}) - \varepsilon/2] \\
&\geq P(xZ_{k-1} \in N_z)\ [P(Z_n \in z^{-1}\tilde{N}_y \text{ f.o.}) - \varepsilon/2] > 0
\end{aligned}
$$

This is a contradiction. Q.E.D.

3.11 Proposition: If $x \to y$ i.o., then xD is a closed right group and D is completely simple.

Proof: (i) By Proposition (3.10), $y \in \bigcap_{z \in xD} \overline{zD} \equiv I$. We claim that I is a minimal right ideal of D and hence by Lemma (3.6), D is completely simple. We only need to prove that I is right simple, i.e., $sI = I$ for all $s \in I$. Let $s \in I$ and let $w \in sI \subseteq I \subseteq \overline{xD} \subseteq xS$ = a right group. Then sI is a right ideal of D and $\overline{sI} = sI$, since the left translations in a right group are closed. Hence, $I \subseteq \overline{wD} \subseteq sI \subseteq I$. This completes the proof.

Under the hypothesis of Proposition (3.11), D becomes also completely simple. In such a case, in the representation of $S = E \times G \times F$ we may (and do) choose:

$$E = E(Se) \supseteq E' \equiv E(De), \text{ where } e \text{ is some element in } E(D)$$
$$(\equiv \text{the set of idempotents in } D).$$
$$F = E(eS) \supseteq F' \equiv E(eD)$$
$$G \equiv eSe \supseteq G' \equiv eDe, \text{ so that}$$
$$D = E' \times G' \times F' \subseteq S = E \times G \times F.$$

This representation will be used in the proof of Theorem (3.15) in the sequel.

3.12 Definition: We write $G_{ef} = \{e\} \times G \times \{f\}$ = a typical maximal subgroup of $S = E \times G \times F$, and $u_{ef} \equiv (e, (fe)^{-1}, f)$ = the unit of G_{ef}.

The ensuing five theorems give recurrence criteria similar to those available for groups [cf. (3.1)]. It is shown that a state is conservative if and only if it is recurrent. Either $R^r = \phi$ or R^r is the completely simple minimal ideal of D and coincides with D. They also give a criterion for $S = E \times G \times F$ to support a recurrent random walk.

3.13 Theorem: Let $S = E \times G \times F$. Then for each idempotent $u_{ef} \in S$, $u_{ef} \to u_{ef}$ i.o. if and only if $\sum_{n=1}^{\infty} P_{u_{ef}}(Z_n \in N) = \infty$ for every neighborhood N of u_{ef}.

Proof: The "$\Rightarrow$" part is trivial by the Borel-Cantelli lemma. We shall prove the "$\Leftarrow$" part in four steps. In Step I, we show that $P(Z_n \notin N^{-1}N$ for all $n \geq k) = 0$. But transition from $N^{-1}N$ to $u_{ef}^{-1}N$ involves complications and this is demonstrated in Steps II and III. Then Step IV completes the proof.

Step I. Let k be a fixed positive integer. Let N be a given neighborhood of u_{ef}. Since

$$\infty = \sum_{n=1}^{\infty} P_{u_{ef}}(Z_n \in N) = \sum_{j=1}^{k} \sum_{i=0}^{\infty} P_{u_{ef}}(Z_{j+ik} \in N),$$

there is a j_0 such that $1 \leq j_0 \leq k$ and $\sum_{i=0}^{\infty} P_{u_{ef}}(Z_{j_0+ik} \in N) = \infty$. Now,

$$1 \geq P_{u_{ef}}(Z_n \in N \text{ for finitely many } n)$$

$$\geq \sum_{i=0}^{\infty} P_{u_{ef}}(Z_{j_0+ik} \in N,\ Z_n \notin N \text{ for all } n \geq j_0 + (i+1)k)$$

(these sets are pairwise disjoint)

(3)

$$\geq \sum_{i=0}^{\infty} P_{u_{ef}}(Z_{j_0+ik} \in N,\ Z_{j_0+ik}^{-1} Z_n \notin N^{-1}N \text{ for all } n \geq j_0 + (i+1)k)$$

$$= \sum_{i=0}^{\infty} P_{u_{ef}}(Z_{j_0+ik} \in N)\, P(Z_n \notin N^{-1}N \text{ for all } n \geq k).$$

Hence,

(4) $$P(Z_n \notin N^{-1}N \text{ for all } n \geq k) = 0.$$

(This result is true for any arbitrary neighborhood N′ in $\{e\} \times G \times F$ of u_{ef}).

Step II. We find $N_0 \subset N$ such that $N_0 = \{(a,b,c);\ c \in N_f,\ b \in N_u(ce)^{-1}\}$, where N_f is a neighborhood of f and N_u is a neighborhood of the identity u in the group G. Then it can be checked easily that

$$u_{ef}^{-1} N_0 = \{(e',b,c);\ b \in (fe')^{-1}(fe)\, N_u(ce)^{-1},\ c \in N_f,\ e' \in E\}.$$

Let E_1 be a compact subset of E.

We can find neighborhoods $N_f' \subset N_f$, $N_u' \subset N_u$ of f and u respectively such that

$$N_1 = \{(e', b, e);\ c \in N_f',\ b \in N_u'\ (ce')^{-1},\ e' \in E_1\} \subset u_{ef}^{-1} N_0$$

(see computation of $u_{ef}^{-1} N_0$ above). This is possible because for every $e' \in E_1$ (which is compact), $(fe)^{-1}(fe')u(fe')^{-1}(fe) \subset N_u$. (Note that the mapping $(f,e) \to f\cdot e$ from $F \times E \to G$ is continuous by the definition of a completely simple semigroup). Hence we can find N_f', N_u' such that for every $e' \in E_1$ and, for every $c \in N_f'$, $(fe)^{-1}(fe')N_u'\ (ce')^{-1}(ce) \subset N_u$, or $N_u'(ce')^{-1} \subset (fe')^{-1}(fe)N_u(ce)^{-1}$,

for every $c \in N_f'$ and every $e' \in E_1$.

Now we can find $N_f''(\subset N_f')$, $N_u''(\subset N_u')$ such that

$$N_2 = \{(e',b,c);\ c \in N_f'',\ b \in N_u''(ce')^{-1},\ e' \in E_1\} \subset N_1,$$

and

$$N_2^{-1}N_2 \cap E_1 \times G \times F \subset N_1.$$

This is possible since

$N_2^{-1}N_2 \cap E_1 \times G \times F$

$$=\{(e',b,c);(e'',b',c')(e',b,c) \in N_2,\text{ where } (e'',b',c') \in N_2, e' \in E_1\}$$

$$=\{(e',b,c);\ e' \in E_1, b'(c'e')b \in N_u''\ (ce'')^{-1},$$

$$\text{where } e'' \in E_1,\ c \in N_f'',\ c' \in N_f'', b' \in N_u''(c'e'')^{-1}\}$$

$$\{(e',b,c);\ e' \in E_1, b \in (c'e')^{-1}(c'e'')N_u''^{-1}N_u''(ce'')^{-1},$$

$$\text{where } c' \in N_f'',\ e'' \in E_1, c \in N_f''\}$$

which can be seen to be a subset of N_1 by properly choosing N_f'' and N_u'' since for every e', $e'' \in E_1$, $(fe')^{-1}(fe'')u(fe'')^{-1}(fe') \subset N_u'$, so that there exist N_f'', N_u'' such that

$$(N_f''e')^{-1}(N_f''e'')N_u''^{-1}N_u''(N_f''e'')^{-1}(N_f''e') \subset N_u' \ .$$

Step III. We claim that for all k (a positive integer)

$$P_{u_{ef}}(Z_n \notin N \text{ for all } n \geq k) = 0 .$$

To prove this, let $\varepsilon > 0$. Let E_1 be a compact subset of E such that $\mu((E - E_1)\times G \times F) < \varepsilon$. By Step II, we can find N_2, a relative neighborhood of u_{ef} in $\{e\}\times G \times F$, such that $N_2^{-1}N_2 \cap E_1 \times G \times F \subset u_{ef}^{-1}N$. Now,

$$\begin{aligned}
P_{u_{ef}}(Z_n \notin N \text{ all } n \geq K) &\leq P(Z_n \notin u_{ef}^{-1} N \text{ all } n \geq k) \\
&\leq P(Z_n \notin N_2^{-1}N_2 \cap E_1 \times G \times F \text{ all } n \geq k) \\
&\leq P(Z_n \notin N_2^{-1}N_2 \text{ all } n \geq k) \\
&\quad + P(Z_n \in (E - E_1) \times G \times F \text{ for some } n) \\
&= 0 + \varepsilon \text{ (by Step I)}.
\end{aligned}$$

(Note that $\{Z_n \in (E - E_1) \times G \times F$ for some $n\}$ is contained in $\{Z_1 \in (E - E_1) \times G \times F\}$.)

Since $\varepsilon > 0$ is arbitrary, our claim is proven.

Step IV. We have

$$\begin{aligned}
&P_{u_{ef}}(Z_n \in N \text{ for finitely many } n) \\
&= \sum_{k=1}^{\infty} P_{u_{ef}}(Z_k \in N, Z_n \notin N \text{ for all } n \geq k + 1) \\
&\leq \sum_{k=1}^{\infty} P_{u_{ef}}(Z_n \notin N \text{ for all } n \geq k + 1) = 0.
\end{aligned}$$

Hence, $P_{u_{ef}}(Z_n \in N \text{ i.o.}) = 1$. Q.E.D.

3.14 Proposition: If $x \to y$ i.o. for some x,y, then $xD \subset R^r$.

Proof: By Proposition (3.11), xD is a closed right group. Let $E(xD)$ be the set of idempotents of xD and let $f \in E(xD)$. Since $y \to f$, for each neighborhood N of f there is k such that $P_y(Z_k \in N) > 0$; by lower semicontinuity in x of the function $\mu^k(x^{-1}N)$, there is a neighborhood of y, N_y, such that $P_{x'}(Z_k \in N) > \delta > 0$, for all $x' \in N_y$. Now for $n > k$,

$$P_f(Z_n \in N) = \int P_f(Z_{n-k} \in dx')P_{x'}(Z_{n-k}^{-1}Z_n \in N) \geq \delta\, P_f(Z_{n-k} \in N_y).$$

By Proposition (3.10), $f \to y$ i.o., so $\sum P_f(Z_n \in N_y) = \infty$ by the Borel-Cantelli lemma. Hence by Theorem (3.13) and the above inequality, $f \in R$ for every $f \in E(xD)$. Since xD is a right group, $xD = \cup\{xDf;\ \text{where } f \in E(xD)\}$. Since R is a left ideal of D (Lemma 3.7), we have $xD \subset R$. Q.E.D.

3.15 Theorem: (i) If $x \to y$ i.o. for some x,y, then $R^r = D$ = a completely simple semigroup and for every pair $z,w \in aD$, a = any element of D, we have $w \to z$ i.o.

(ii) Either $R^r = \phi$ or $R^r = D$ = a completely simple semigroup.

Proof: By Proposition (3.11) $D = E \times G \times F$ is completely simple and by Proposition (3.14) there is an idempotent $u_{ef} = (e,(fe)^{-1},f) \in R$. Since $(e',g,f)\,(e,(fe)^{-1},f) = (e',g,f) \in R$ (since R is a left ideal of D by Lemma (3.7), we obtain $G_{e'f} \equiv \{e'\} \times G \times \{f\} \subset R$ for all $e' \in E$. Since $(e,(fe)^{-1},f)(e,g,f') = (e,g,f') \subset u_{ef}D \subset R$ by Prop. (3.14), $G_{ef'} \subset R$ for every $f' \in F$. Consider next $G_{e'f'}$, for any $e' \in E$, $f' \in F$. Since the idempotent $(e',(fe')^{-1},f) \in R$, again by the above argument, $G_{e'f'} \subset R$ and hence $R = D$. The rest of the claims follow easily from Proposition (3.10). Q.E.D.

3.16 Theorem: $R^r \neq \phi$ if and only if $\sum_{n=1}^{\infty} P(Z_n \in N) = \infty$ for all open neighborhoods N of some $x \in D$.

Proof: If $R^r \neq \phi$, then $R^r = D$ by Theorem (3.15). Hence $\sum P_y(Z_n \in N) = \infty$ for all $y \in D \cap N$. Choose k such that $\mu^k(D \cap N) > 0$. Then,

$$\sum_{n=1}^{\infty} P(Z_n \in N) \geq \sum_{n=k+1}^{\infty} \int_D P_y(Z_{n-k} \in N)\mu^k(dy)$$

$$= \int \sum_{n=k+1}^{\infty} P_y(Z_{n-k} \in N)\mu^k(dy) = \infty .$$

We prove next the "<==" part: Let $\sum_{n=1}^{\infty} P(Z_n \in N) = \infty$ for all neighborhoods N of some $x' \in D$. Then for each neighborhood N of $x' = (e,g,f)$, we have as in the proof of Theorem (3.13) for every positive integer k,

$$P(Z_n \notin N^{-1}N \text{ for all } n \geq k) = 0.$$

Now given an open neighborhood $N_1 = N_0(e) \times N_0((fe)^{-1}) \times N_0(f)$ of $(e,(fe)^{-1},f)$, we can find an open neighborhood $N = N_0(e) \times g(fe)N_0((fe)^{-1} \times N_0(f)$ of $x' = (e,g,f)$ such that $N^{-1}N = N_1^{-1}N_1$. The reason is as follows: Let $(x,y,z) \in N^{-1}N$. Then there exist $e' \in N_0(e), g' \in N_0((fe)^{-1})$, $f' \in N_0(f)$ such that

$$(e',g(fe)g',f')(x,y,z) \in N \quad \text{or} \quad (e',g(fe)g'(f'x)y,z) \in N.$$

This means that $g'(f'x)y \in N_0((fe)^{-1})$ or $(e',g',f')(x,y,z) \in N_1$ or $(x,y,z) \in N_1^{-1}N_1$. Hence $N^{-1}N \subset N_1^{-1}N_1$. Similarly $N_1^{-1}N_1 \subset N^{-1}N$. Hence for every positive integer k, $P(Z_n \notin N_1^{-1}N_1$ for all $n \geq k) = 0$, for every open neighborhood N_1 of $(e,(fe)^{-1},f)$. Now by following the proofs of Steps II, III, and IV in Theorem (3.13), we can prove that for every neighborhood N_1 of $(e,(fe)^{-1},f) \equiv u_{ef}$,

$$P_{u_{ef}}(Z_n \in N_1 \text{ i.o.}) = 1.$$

Hence $u_{ef} \in R$ and $R \neq \phi$. Q.E.D.

3.17 <u>Corollary</u>: $R^r \neq \phi$ if and only if $\sum_{n=1}^{\infty} P_x(Z_n \in N) = \infty$ for all neighborhoods N of some $y \in D$, x being some fixed element of D.

Observe that $\sum P_x(Z_n \in N_y) = \infty$ for all N_y <u>implies</u> that $y \in \overline{xD} \subset \{e\} \times G \times F$, where $x = (e,g,f)$. Also equations (3), (4) in Th. (3.13) are valid with $P_{u_{ef}}$ replaced by P_x - measure and the last part of the proof of Theorem (3.16) applies in this case.

3.18 <u>Note:</u> From Theorems (3.15) and (3.16), one easily sees that if the right walk is recurrent ($R^r \neq \phi$), so is the left ($R^\ell \neq \phi$) and conversely.

If in particular S = a right group and $R^r \neq \phi$, then $R_u^r = R^r =$

$R^\ell = D$ = a right subgroup of S but R_u^ℓ may be empty, where R_u^ℓ, R^ℓ are the corresponding recurrence sets for the left random walk. For example, let $S = G \times E$, a compact right group and μ be an idempotent measure whose support is all of S. We choose E to consist of more than one point. Then if the left walk had an unconditionally recurrent state ($R_u^\ell \neq \phi$), from Theorem (3.4), S would be a group and this contradicts the fact that E is not a singleton. Of course, every element of S is recurrent for the right walk Z_n.

3.19 Theorem: Suppose that the group factor G is compact in the representation of $S = E \times G \times F$. Then Z_n is always recurrent and (as in Th. (3.15)) $R^r = D$ = a completely simple subsemigroup.

Proof: Let J and H be compact subsets of E and F such that $\mu(J \times G \times H) > 0$. (This can be done by the regularity of μ.) Let $L = E \times G \times H$ and $L_1 = J \times G \times H$; then for any $x \in L_1 \cap D$,

$$P_x(Z_n \in L_1 \text{ i.o.}) = \lim_{k\to\infty} P_x\left(\cup_{n=k}^{\infty}\{Z_n \in L\}\right)$$

$$= \lim_{k\to\infty} P_x\left(\cup_{n=k}^{\infty}\{X_n \in L\}\right)$$

$$= \lim_{k\to\infty}\left[1 - P_x\left(\cap_{n=k}^{\infty}\{X_n \notin L\}\right)\right]$$

$$= \lim_{k\to\infty}\left[1 - \prod_{n=k}^{\infty} P_x\{X_n \notin L\}\right] = 1,$$

because $P(X_n \notin L) = \mu(S - L) < 1$, for all n. Hence for each subset $J \subset E$ such that $\mu(J \times G \times H) > 0$, we have

$$P(Z_n \in J \times G \times H \text{ i.o.}) = \int_{(E\times G\times F)\cap D} P_x(Z_n \in J \times G \times H \text{ i.o.})\mu(dx)$$

$$\geq \int_{(J\times G\times H)\cap D} P_x(Z_n \in J \times G \times H \text{ i.o.})\mu(dx)$$

$$= \mu(J \times G \times H) > 0.$$

Now suppose that Z_n is not recurrent. For every $x \in D \cap J \times G \times H$, there exists a neighborhood $N(x)$ such that $\sum P(Z_n \in N(x)) < \infty$, by Theorem (3.16). By the Borel-Cantelli lemma, $P(Z_n \in N(x) \text{ i.o.}) = 0$. Since $J \times G \times H$ is compact, we can find $x_1, x_2, \ldots x_m \in J \times G \times H$ and neighborhoods $N(x_1), N(x_2), \ldots, N(x_m)$ such that

$$J \times G \times H \subset \cup_{i=1}^{m} N(x_i)$$

and

$$P(Z_n \in J \times G \times H \text{ i.o.}) \leq \sum_{k=1}^{m} P(Z_n \in N(x_k) \text{ i.o.}) = 0$$

which is a contradiction. Q.E.D.

4. The three random walks in the case of compact semigroups.

In this section we will study the three random walks on compact semigroups. It turns out that the compact case is one of the interesting cases where information concerning the recurrence of all the three random walks is fairly complete.

It will be shown that the unilateral and bilateral walks are recurrent and that the set of their recurrent states is precisely the minimal ideal (= kernel) K of S. ($R^r = R^\ell = R^b = K$). Moreover, conservative, recurrent and essential states coincide and all recurrent states are recurrent positive in the three random walks. Also the structure of the essential classes for the bilateral walk (which are at most two) will be given. These results were obtained by the authors in [18] and [19]. Larisse introduced in [14] the bilateral walk W_n and proved (under the hypothesis that one of the walks is recurrent) similar results for the bilateral walk when D is discrete (countable).

It will be seen that the methods used in the bilateral walk are generally different from those in the case of unilateral walks, the reason being the different form of the n - transition function for W_n.

Throughout this section S will be compact. We shall use the notations

$$y^{-1}B = \{x \in S : yx \in B\}$$

$$A^{-1}B = \cup \{y^{-1}B : y \in A\}$$

$$Z_k^{-1}Z_n = X_{k+1}X_{k+2} \cdots X_n$$

$$\mu^n * \mu^n(x^{-1} \cdot) = \mu^n *_x(\mu^n) = P_b^n (x, \cdot)$$

$$K = \text{the minimal ideal } (= \text{kernel}) \text{ of } D.$$

We shall consider first the unilateral walk. Before we present the main results, we need the following lemmas.

4.1 Lemma: Given $x \in S$ and N_y, an open set containing y in S, we can find open sets N_1 and N_2 containing x and y respectively such that $N_1^{-1} N_2 \subset x^{-1}N_y$.

Proof: Suppose first that $x^{-1}N_y$ is empty. Then $y \notin xS$. If for every open set N' containing y and open set N" containing x, N' $\cap$ N" S is non-empty, then we can find a net y_u converging to y and a net x_u converging to x such that $y_u = x_u s_u$, where s_u's are elements in S. Since S is compact, we can find a subnet of s_u converging to some element s in S. But then y = xs and this contradicts that $x^{-1}N_y$ is empty. Hence, when $x^{-1}N_y$ is empty, there exist open sets N_1 and N_2 containing x and y respectively such that $N_2 \cap N_1S$ is empty which means that $N_1^{-1}N_2$ is empty.

Suppose next that $x^{-1}N_y$ is non-empty. Suppose, for every open set N' containing y and contained in N_0, a compact neighborhood of y such that $N_0 \subset N_y$, and every open set N" containing x, we have $N''^{-1}N' \cap (x^{-1}N_y)^c$ is non-empty. Then we can find a net x_u converging to x and $z_u \in N''^{-1}N' \cap (x^{-1}N_y)^c$ such that $x_u z_u \in N_0$ and $xz_u \notin N_y$. Since S is compact, we can find a subnet of z_u converging to z such that $xz \in N_0$ and $xz \notin N_y$. This is a contradiction and the lemma follows.

4.2 Lemma: If $x \to y$ i.o. and $x \to z$, then $z \to y$ i.o.

Proof: If $z \not\to y$ i.o., then there is an open set N containing y such that $P_z(Z_n \in N \text{ finitely often}) > 0$. By Lemma (4.1), we can find open sets N_z and N_y containing z and y respectively such that $N_z^{-1}N_y \subset z^{-1}N$. But then $P(Z_n \in N_z^{-1}N_y \text{ finitely often}) > 0$. Also

$$(5)\quad P_x(Z_n \in N_y \text{ finitely often}) \geq P_x(Z_k \in N_z, Z_n \in N_y \text{ finitely often})$$

$$\geq P_x(Z_k \in N_z, Z_k^{-1}Z_n \in N_z^{-1}N_y \text{ finitely often})$$

$$= P_x(Z_k \in N_z)P(Z_n \in N_z^{-1}N_y \text{ finitely often})$$

$$\geq P_x(Z_k \in N_z)P(Z_n \in z^{-1}N \text{ f.o.}) > 0,$$

if Z_k is such that $P_x(Z_k \in N_z) > 0$. This is a contradiction to the fact that $x \to y$ i.o. and the lemma follows.

4.3 Lemma: The set R^r of the recurrent states of Z_n is a subset of the kernel K of D.

Proof: Let $x \in R$. Then $x \in xD$. We claim that xD is a minimal right ideal of D. To prove this, let $a \in xD$. Suppose that $x \notin aD$. Then, by Lemma (4.1), we can find open sets N_x and N_a containing x and a respectively such that $N_a^{-1}N_x \cap D$ is empty. Since $a \in xD$, for some positive integer k, $P_x(Z_k \in N_a) > 0$. Now

$$P_x(Z_n \in N_x \text{ finitely often}) \geq P_x(Z_k \in N_a, Z_{n+k} \in N_x \text{ finitely often})$$

$\geq P_x(Z_k \in N_a, Z_k^{-1}Z_{n+k} \in N_a^{-1}N_x$ finitely often)

$= P_x(Z_k \in N_a) > 0.$

This contradicts that $x \in R$ and therefore, $x \in aD$ so that xD is a minimal right ideal of D. Since the kernel K of D is the union of all minimal right ideals of D, $x \in K$.

4.4 Note: It is easy to see that x is an essential state for Z_n if and only if $x \in K$. For if x is essential (since $x \to xk$, for $k \in K$), $xk \to x$ or $x \in xkD \subset xK \subset K$. If $x \in K$ and $x \to y$ then $y \in xD = xkxD \subset xK$ for some $k \in K$. Thus, $y \in xK \cap yK$ and so $y \in xK = yK$ and $y \to x$. (observe that $K = E \times G \times F$ (completely simple) and xK is a minimal right ideal).

4.5 Theorem: Let $y \in D$. Then the following are equivalent:

(a) $y \to y$ i.o. with respect to Z_n.

(b) $\sum_{n=1}^{\infty} \mu^n(y^{-1}N_y) = \infty$ for every open set N_y containing y.

(c) $\sum_{n=1}^{\infty} \mu^n(N_y) = \infty$ for every open set N_y containing y.

(d) $y \to y$ i.o. with respect to L_n (the left random walk).

(e) $\sum_{n=1}^{\infty} \mu^n(N_y y^{-1}) = \infty$ for every open set N_y containing y.

Proof: We will prove that (a), (b) and (c) are equivalent. Then by dual arguments, (c), (d) and (e) also will be equivalent.

First, (a) implies (b), by the Borel-Cantelli Lemma.

Next, we show that (b) implies (c). We assume (b) and that there is an open set N_y containing y such that $\sum_{n=1}^{\infty} \mu^n(N_y) < \infty$. Let N' be a compact neighborhood of y such that $N' \subset N_y$. Then $y^{-1}N'$ is compact. Noting that for $x \in y^{-1}N'$, $yx \in N'$ and so we can find open sets $N_x(y)$ and $N(x)$ containing y and x respectively such that $N_x(y)N(x) \subset N_y$, we can find (using compactness) an open set N containing y such that

$y^{-1}N \subset z^{-1}N_y$ for every z in N. Now for every positive integer k,

$$\sum_{n=1}^{\infty} \mu^n(N_y) \geq \int \sum_{n=k+1}^{\infty} \mu^{n-k}(z^{-1}N_y)\, \mu^k(dz)$$

which means that we can find a set A such that, for every positive integer k, $\mu^k(A) = 0$ and, for $z \notin A$, $\sum_{n=1}^{\infty} \mu^n(z^{-1}N_y) < \infty$. Hence there exists $w \in N$ such that $\sum_{n=1}^{\infty} \mu^n(w^{-1}N_y) < \infty$. Since $y^{-1}N \subset w^{-1}N_y$, this contradicts the assumption of (b).

Finally, we show that (c) implies (a). Let k be an arbitrary positive integer and N_y be an open set containing y. Then if (c) holds, we have $\sum_{j=1}^{k} \sum_{i=0}^{\infty} P(Z_{j+ik} \in N_y) = \infty$. So we can find an integer m such that $1 \leq m \leq k$ and $\sum_{i=0}^{\infty} P(Z_{m+ik} \in N_y) = \infty$. Now

$$\begin{aligned}
1 &\geq P(Z_n \in N_y \text{ finitely often}) \\
&\geq \sum_{i=0}^{\infty} P(Z_{m+ik} \in N_y,\ Z_n \notin N_y \text{ for all } n \geq m + (i+1)k) \\
&\geq \sum_{i=0}^{\infty} P(Z_{m+ik} \in N_y, Z_{m+ik}Z_n \notin N_y^{-1}N_y \text{ for all } n \geq m + (i+1)k) \\
&= P(Z_n \notin N_y^{-1}N_y \text{ for all } n \geq k) \sum_{i=0}^{\infty} P(Z_{m+ik} \in N_y)
\end{aligned}$$

which means that, for any open set N_y containing y and every positive integer k,

$$P(Z_n \notin N_y^{-1}N_y \text{ for all } n \geq k) = 0.$$

Now given an open set N containing y, by Lemma (4.1), we can find an open set N_y containing y such that $N_y^{-1}N_y \subset y^{-1}N_y$. Then

$$P_y(Z_n \in N \text{ finitely often}) \leq P(Z_n \in N_y^{-1}N_y \text{ finitely often}) = 0.$$

Hence $y \to y$ i.o. and (a) holds.

4.6 <u>Theorem</u>: Let $y \in D$ and $x \in K$. Then the following are equivalent:

(a) $x \to y$ i.o. with respect to Z_n.

(b) $\sum_{n=1}^{\infty} \mu^n(x^{-1}N_y) = \infty$ for every open set N_y containing y.

(c) $x \to y$ and $\sum_{n=1}^{\infty} \mu^n(N_y) = \infty$ for every open set N_y containing y.

Proof: First, (a) implies (b), by the Borel-Cantelli Lemma.

Next, we show that (b) implies (c). Suppose that (b) holds. Then by following the proof of "(c) implies (a)" in Theorem (4.5) and noting that

$$P_x(Z_{m+ik} \in N_y,\ Z_{m+ik}^{-1}Z_n \notin N_y^{-1}N_y \text{ for all } n \geq m + (i+1)k)$$

$$= P_x(Z_{m+ik} \in N_y) \cdot P(Z_n \notin N_y^{-1}N_y \text{ for all } n \geq k),$$

we see that, for every open set N_y containing y and each positive integer k, $P(Z_n \notin N_y^{-1}N_y$ for all $n \geq k) = 0$. Then it follows as in Theorem (4.5) that $y \to y$ i.o. This implies (c), by Theorem (4.5).

Finally, we show that (c) implies (a). Since (c) holds, by Theorem (2.4), $y \to y$ i.o. Since $x \to y$, $y \in xD$. But xD is a minimal right ideal of D, x being an element of K. Therefore, $x \in xD = yD$ so that $y \to x$. By Lemma (4.2), $x \to y$ i.o. The proof of the theorem is complete.

Remark. Using Theorem (4.5), it is now easy to show that the set R^r of recurrent states of $\{Z_n\}$ is precisely the kernel K of D. If, for each $y \in D$, y is not recurrent, then by Theorem (4.5), we can find an open neighborhood N_y for each y such that $\sum_{n=1}^{\infty} \mu^n(N_y)$ is finite. But then since D is compact, $\sum_{n=1}^{\infty}\mu^n(D)$ is finite, which is absurd. Hence R^r is nonempty. Also, R^r is a left ideal of D. If R^ℓ is the set of recurrent states of the left random walk $\{L_n\}$, then, by Theorem 4.5 $R^r = R^\ell$. Also, R^ℓ is a right ideal of D. Therefore, R^r is a two-sided ideal of D and hence contains K. By Lemma (4.3), $R^r \subset K$. Hence $R^r = R^\ell = K$. A direct proof of this interesting fact can also be given independently of Theorem (4.5). This is shown in what follows.

4.7 Theorem: The set R^r of recurrent states of Z_n is precisely the kernel K of D.

Proof: Because of Lemma (4.3), suffice it to show that $K \subset R$. We will use the fact that for any regular probability measure Q on S, the function $x \to Q(x^{-1}U)$ is lower semicontinuous for open U. Now let $x \in K$. For any $y \in xD$, we have $x \in yD$. For any open neighborhood U of x, $U \cap yD$ is nonempty. Thus for some n, $U \cap yS_{\mu}^{n}$ is nonempty. Let

$$U_n = \{y \in xD : U \cap yS_{\mu}^{n} \text{ is nonempty}\}.$$

Then U_n is relatively open in the compact set xD. So there is a finite subcover and, for some finite n,

$$g(y) \equiv \sum_{j \leq n} \mu^j (y^{-1}U) > 0$$

for all $y \in xD$. Then by the lower semicontinuity of g(y), there is a $p > 0$ such that $g(y) \geq p$ for all $y \in xD$. Let $q = p/n$. For any positive integer k, the conditional probability that $xZ_m \notin U$ for $nk < m \leq n(k+1)$, given any conditions on X_j for $j \leq nk$, is at most $1 - q$. Hence

$$P(xZ_m \notin U \text{ for } nk < m \leq nN) \leq (1 - q)^{N-k}$$

which converges to zero as N approaches infinity. This means that

$$P_x(Z_m \in U \text{ finitely often}) = 0$$

so that $x \in R$. The proof is complete.

From Theorems (4.5) and (4.6) and from the proof (part (c) ==> (a)) of (4.5), we can obtain the following neat criterion for recurrence.

4.8 Theorem: A state $y \in D$ is recurrent in the unilateral walks (and hence is in the kernel K) if and only if $\sum_n \mu^n(N(y)x^{-1}) = \infty$ for all open neighborhoods N(y) of y and some state $x \in D$.

Proof: We let $Z_n = X_1X_2 \ldots X_n$ and $Z_k^{-1}Z_n = X_{k+1}X_{k+2} \ldots X_n$. In case y is recurrent, then $P_x(Z_n \in N(y) \text{ i.o.}) = 1$ for every x in the recurrence class yD. Next, suppose $\sum \mu^n(N(y)x^{-1}) = \infty$ for all N(y) and some $x \in D$. Let k be arbitrary positive integer. Then

$$\sum_{j=1}^{k} \sum_{i=0}^{\infty} P_x(Z_{m+ik} \in N(y)) = \infty$$

So we can find an integer m such that $1 \le m \le k$ and $\sum_{i=0}^{\infty} P_x(Z_{m+ik} \in N(y)) = \infty$.

Now

$$1 \geq P_x(Z_n \in N(y) \text{ finitely often})$$
$$\geq \sum_{i=0}^{\infty} P_x(Z_{m+ik} \in N_y,\ Z_n \notin N_y \text{ for all } n \geq m + (i+1)k)$$
$$\geq \sum_{i=0}^{\infty} P_x(Z_{m+ik} \in N_y,\ Z_{m+ik}^{-1} Z_n \notin N_y^{-1} N_y \text{ for all } n \geq m + (i+1)k)$$
$$= P(Z_n \notin N_y^{-1} N_y \text{ for all } n \geq k) \sum_{i=0}^{\infty} P_x(Z_{m+ik} \in N(y))$$

which means that, for any open set $N_y = N(y)$ containing y and every positive integer k,

$$P(Z_n \notin N_y^{-1} N_y \text{ for all } n \geq k) = 0.$$

(Here $N_y^{-1} N_y \equiv \{s;\ zs \in N_y \text{ for some } z \in N_y\} = \bigcup \{z^{-1} N_y; z \in N_y\}$.) Now given an open set N containing y, by Lemma (4.1), we can find an open set N_y containing y such that $N_y^{-1} N_y \subset y^{-1} N$. Then

$$\begin{aligned} P_y(Z_n \in N \text{ finitely often}) &\leq P(Z_n \in y^{-1} N \text{ f.o.}) \leq P(Z_n \in N_y^{-1} N_y \text{ f.o.}) \\ &= \sum_{i=1}^{\infty} P(Z_i \in N_y^{-1} N_y, Z_n \notin N_y^{-1} N_y \text{ for all } n > i) = 0. \end{aligned}$$

Hence $P_y(Z_n \in N \text{ i.o.}) = 1$ and y is recurrent and belongs to K.

4.9 <u>Proposition:</u> For any neighborhood U of the kernel K and any $x \in D$

$$P(Z_n \in U \text{ i.o.}) = P_x(Z_n \in U \text{ i.o.}) = 1$$

<u>Proof:</u> There exists an open ideal I such that $K \subset I \subset U$ by a result of R. J. Koch and A. D. Wallace [13]. There is k such that $\mu^k(I) > 0$. We observe that

$$\mu^{n+k}(I^c) = \int_{I^c} \mu^n(I^c x^{-1}) \mu^k(dx) \leq \mu^n(I^c) \mu^k(I^c); \quad I^c = D - I.$$

Also,

$$\sum\mu^n(I^c) = (\text{Constant}) + \mu^k(I^c) + \ldots + \mu^{2k-1}(I^c) + \mu^{2k}(I^c) + \ldots$$
$$+ \ldots \mu^{3k-1}(I^c) + \ldots$$
$$\leq (\text{Constant}) + k\mu^k(I^c) + k[\mu^k(I^c)]^2 + k[\mu^k(I^c)]^3 + \ldots < \infty.$$

Hence by the Borel-Cantelli lemma, $P(Z_n \in I^c \text{ i.o.}) = 0$ and $P(Z_n \in I \text{ i.o.}) = 1$.

(Observe that also $\sum\mu^n(I^c x^{-1}) \leq \sum\mu^n(I^c) < \infty$ for every $x \in D$.) Q.E.D.

In the remaining of this Section we consider the bilateral walk $W_n = X_{-n}\ldots X_{-1}X_0X_1\ldots X_n$ on an arbitrary compact semigroup S.

4.10 Lemma: Let B be a Borel subset of D. Then $A = \{(y,z) : yxz \in B, y, z \in D\}$ is (Borel) measurable and

$$P_b^n(x,B) \equiv \mu^n \times \mu^n (A) = \mu^n * \mu^n(x^{-1}\cdot)(B) \equiv \mu^n *_x(\mu^n)(B).$$

Proof: If $\psi(y,z) = yxz$, then ψ is a continuous mapping from $D \times D$ into D so that $A = \psi^{-1}(B)$ is also measurable in $D \times D$ (By 2nd countability). Let A_y = the section of A by y. By Fubini's theorem,

$$\mu^n \times \mu^n(A) = \int \mu^n(A_y)\mu^n(dy) = \int \mu^n(x^{-1}y^{-1}B)\mu^n(dy) = \mu^n *_x(\mu^n)(B).$$

4.11 Lemma: Let U be an open set in D. Then the mapping $x \to \mu^n *_x(\mu^n)(U)$ is lower semicontinuous.

Proof: Let $y_k \to y$. Then by Fatou's theorem,

$$\underline{\lim}_{k\to\infty} \int \mu^n(y_k^{-1}z^{-1}U)\,\mu^n(dz) \geq \int \underline{\lim}\,\mu^n(y_k^{-1}z^{-1}U)\mu^n(dz)$$
$$\geq \int \mu^n(y^{-1}z^{-1}U)\mu^n(dz)$$

since $y \to \mu^n(y^{-1}z^{-1}U)$ is a lower semicontinuous function by (2.6).

4.12 Lemma: If $x \in K$ (= the kernel of D), then $x \to x$ (with respect to the bilateral walk).

Proof. Let $x \in K$. Since K is completely simple, $xKx = xDx$ is a group and $x \in xKx$. Let e be the identity of xKx and let $y \in xKx$ such that $xy = yx = e$. Since $xyex = x$, given a neighborhood N(x) of x, there exists $N_1(e)$ such that $xyN_1(e)N_1(e)x \subset N(x)$. Let F stand for S_μ. Since $e \in xD$, $N_1(e) \cap xF^k \neq \phi$ for some k and so $xyN_1(e)N_1(e)x \cap xyxF^k xF^k x \neq \phi$. Hence $N(x) \cap (xF^k xF^k x) \neq \phi$. Therefore there is $w \in F^k xF^k$ such that $xwx \in N(x)$. Now there is $N_2(x)$ such that $N_2(x)wN_2(x) \subset N(x)$. Also $N_2(x) \cap F^m \neq \phi$ for some m. Hence $N(x) \cap F^{m+k}xF^{m+k} \neq \phi$. This means that $x \in \overline{\bigcup_{n=1}^\infty F^n xF^n}$ or $x \to x$.

4.13 Lemma. A state x is an essential state (bilateral walk) if and only if $x \in K$.

Proof. First, suppose x is essential. Let F stand for S_μ. Then for all $y \in \overline{\bigcup_{n=1}^\infty F^n xF^n}$, $x \in \overline{\bigcup_{n=1}^\infty F^n yF^n}$. Clearly $x \in DxD$ (= an ideal of D).

Let $z \in DxD = \overline{\bigcup_{n,m} F^n xF^m}$. Then there exist $z_n = f_n x f_{m_n}$ ($f_n \in F^n, f_{m_n} \in F^{m_n}$) such that $z_n \to Z$. If $m > n$, let $f_{m_n - n} \in F^{m_n - n}$. Then $f_{m_n - n} f_n f_{m_n} \in F^{m_n} xF^{m_n}$ and so $x \in \overline{\bigcup_{k=1}^\infty F^k f_{m_n-n} f_n x f_{m_n} F^k} \subset Dz_nD$ for all n. Since D is compact, $x \in DzD$. This means that $DxD \subset DzD \subset DxD$ for all $z \in DxD$ or DxD is the kernel K of D. Hence $x \in K$.

Second, let $w \in K$ = the kernel of D. We now show that w is essential in two steps.

Step I. Let $x, y \in eKe$, e being some idempotent of K (eKe being a group with identity e) such that $xy = e$. Given N(y), there exists $N_1(e)$ such that $N(y) \supset yN_1(e)N_1(e)e$. Now $e \in xD$ so that $N_1(e) \cap xF^n \neq \phi$ for some n or $N(y) \cap yxF^n xF^n e \neq \phi$. This means that $N(y) \cap F^m xF^m \neq \phi$ for some m, or $y \in \overline{\bigcup_{k=1}^\infty F^k xF^k}$, or $x \to y$. Similarly we show that $y \to x$.

Step II. Let $w \in K$. Then if e is the identity of the group wKw, we have $eKe = wKw$. Let $y = f_k w g_k$, $f_k \in F^k$, $g_k \in F^k$. Then $eye =$ $(ef_k e)w(eg_k e) \subset \overline{\bigcup_{n=1}^{\infty} F^n w F^n}$. Now $w = (ef_k e)^{-1}(eye)(eg_k e)^{-1}$ (the inverse being taken in eKe) or $w^{-1} = (eg_k e)(eye)^{-1}(ef_k e)$ which means that $(eye)^{-1} \to w^{-1}$. Now

$$eye \xrightarrow{\quad\quad} (eye)^{-1} \xrightarrow[\text{(by Step I)}]{\quad\quad} w^{-1} \xrightarrow[\text{(by Step I)}]{\quad\quad} w,$$

so that $eye \to w$. This means that $w \in \overline{\bigcup_{n=1}^{\infty} F^n eye\, F^n} \subset \overline{\bigcup_{n=1}^{\infty} F^n y F^n}$ or $y \to w$ (for every $y \in \bigcup_{n=1}^{\infty} F^n w F^n$).

[The reason why $\overline{\bigcup_{n=1}^{\infty} F^n eye F^n} \subset \overline{\bigcup_{n=1}^{\infty} F^n y F^n}$ is as follows: $e \in \overline{\bigcup_k F^k} \Longrightarrow \exists\, z_\alpha \in \bigcup_k F^k$, z_α converging to e so that $z_\alpha y z_\alpha$ converge to eye or $eye \in \overline{\bigcup_k F^k y F^k}$. Hence for all n, $F^n eye F^n \overline{\bigcup_k F^k y F^k F^n} \subset$ $\overline{F^n(\bigcup_k^{\infty} F^k y F^k) F^n} \subset \overline{\bigcup_k^{\infty} F^{k+n} y F^{k+n}} \subset \overline{\bigcup_k^{\infty} F^k y F^k}$].

Now let $z \in \overline{\bigcup_{n=1}^{\infty} F^n w F^n}$. Then there exist $y_\alpha \in \bigcup_{k=1}^{\infty} F^n w F^n$ such that $y_\alpha \to z$ and $w \in \overline{\bigcup_{n=1}^{\infty} F^n y_\alpha F^n}$ for all α. Let N(w) be a compact neighborhood of w. Then $N(w) \cap (\bigcup_{n=1}^{\infty} F^n y F^n) \neq \phi$ for all α. There exist t_α, $s_\alpha \in F^m$ for some m (same m for both t_α, s_α) such that $t_\alpha y_\alpha s_\alpha \in N(w)$. By compactness of D, we can find subnets t_β, y_β, s_β, such that $t_\beta \to t$, $y_\beta \to z$, $s_\beta \to s$ and $tzs \in N(w)$. Since $t_\beta z s_\beta \in \bigcup_{n=1}^{\infty} F^n z F^n$, $tzs \in \overline{\bigcup_{n-1}^{\infty} F^n z F^n}$ so that $N(w) \cap (\overline{\bigcup_{n=1}^{\infty} F^n z F^n}) \neq \phi$. This means that $z \to w$. Q.E.D.

4.14 Theorem: A state x is essential in the bilateral walk if and only if $P_x(W_n \in N(x) \text{ i.o.}) = 1$ for all N(x) of x.

Proof: Suppose $P_x(W_n \in N(x) \text{ i.o.}) = 1$ for all N(x). Let F stand for S_μ. Claim: x is essential and $x \in K$. Suppose $x \notin K$. Then there is

$y \in \overline{\bigcup_{n=1}^\infty F^n x F^n}$ such that $x \notin \overline{\bigcup_{n=1}^\infty F^n y F^n}$. By using compactness of D,

we can find N(x), N(y) such that $N(x) \cap (\bigcup_{n=1}^\infty F^n N(y) F^n) = \emptyset$ and

$N(y) \cap (F^k x F^k) \neq \emptyset$ for some k. Now,

$$0 \leq P(X_{-k} \ldots X_{-1} x X_1 \ldots X_k \in N(y))$$

$$= P(X_{-n} \ldots X_{-1} x X_1 \ldots X_n \in N(x) \text{ i.o.}, X_{-k} \ldots X_{-1} x X_1 \ldots X_k \in N(y))$$

$$= \int_{N(y)} P(X_{-n} \ldots X_{-1} x X_1 \ldots X_n \in N(x) \text{ i.o.} \mid$$

$$X_{-k} \ldots X_{-1} x X_1 \ldots X_k = Z) \mu^k * {}_x(\mu^k)(dz)$$

$$= \int_{N(y)} P(X_{-n} \ldots X_{-1} z X_1 \ldots X_n \in N(x) \text{ i.o.}) \mu^k * {}_x(\mu^k)(dz)$$

$= 0$ since $N(y) \cap (F^n N(y) F^n) = \emptyset$ for all n.

The above contradiction proves that $x \in K$ and therefore x is essential.

Conversely, suppose $x \in K$. Then for any

$y \in \overline{\bigcup_{n=1}^\infty F^n x F^n}$, $x \in \overline{\bigcup_{n=1}^\infty F^n y F^n}$. Let U be an open neighborhood of x. Then $U \cap (F^n y F^n) \neq \emptyset$ for some n. Let $U_n = \{y \in \overline{\bigcup_{n=1}^\infty F^n x F^n};\ U \cap F^n y F^n \neq \emptyset\}$. Clearly the U_n's are relatively open in $\overline{\bigcup_{n=1}^\infty F^n x F^n}$. (For the support of $P_b^n(y,\cdot) = \mu^n * [\mu^n(y^{-1}\cdot)](\cdot)$ is $\overline{F^n y F^n}$. It follows that $U_n = \{y \in \overline{\bigcup F^n x F^n};\ P_b^n(y,U) > 0\}$. But by Lemma (4.11), $P_b^n(\cdot,U)$ is lower semicontinuous so U_n are relatively open.) By compactness, there is a finite subcover and for some finite m,

$$g(y) = \sum_{1\leq j\leq m} \mu^j * {}_y(\mu^j)(U) > 0, \text{ for all } y \in \overline{\bigcup_{n=1}^{\infty} F^n x F^n}.$$

By lower semicontinuity of g(y) (Lemma(4.11)), there is $p > 0$ such that $g(y) \geq p$ for all $y \in \overline{\bigcup_{n=1}^{\infty} F^n x F^n}$. Let $q = p/m$. Now

$$P(X_{-k}\ldots X_{-1}zX_1\ldots X_k \notin U,\ 1 \leq k \leq m) \leq 1-q, \quad \text{for all } z \in \overline{\bigcup_{n=1}^{\infty} F^n x F^n}.$$

Also for any positive integer s,

$$P(X_{-k}\ldots X_{-1}xX_1\ldots X_k \notin U,\ ms < k \leq m(s+1))$$

$$= \int P(X_{-k}\ldots X_{-1}xX_1\ldots X_k \notin U,\ ms < k \leq m(s+1) \mid X_{-ms}\ldots X_{-1}xX_1\ldots X_{ms} = z)\mu^{ms} * {}_x(\mu^{ms})(dz)$$

$$= \int P(X_{-k}\ldots X_{-1}zX_1\ldots X_k \notin U,\ 1 \leq k \leq m)\mu^{ms} * {}_x(\mu^{ms})(dz) \leq 1 - q.$$

Now setting $W_k = X_{-k}..X_{-1}xX_1..X_k$ we have

$$P(X_{-k}\ldots X_{-1}xX_1\ldots X_k \notin U,\ ms < k \leq m(s+2))$$

$$= P(W_k \notin U,\ m(s+1) < k \leq m(s+2) \mid W_k \notin U,\ ms < k \leq m(s+1))$$
$$\times P(W_k \notin U,\ ms < k \leq m(s+1))$$

$$= P(W_k \notin U,\ m(s+1) < k \leq m(s+2) \mid W_{m(s+1)} \notin U)$$
$$\times P(W_k \notin U,\ ms < k \leq m(s+1))$$

$$= \frac{1}{\mu^{m(s+1)} * {}_x(\mu^{m(s+1)})(U^c)} \int_{U^c} P(X_{-k}\ldots X_{-1}zX_1\ldots X_k \notin U,\ 1 \leq k \leq m) \cdot \mu^{m(s+1)} * {}_x(\mu^{m(s+1)})(dz)$$

$$\times P(W_k \notin U,\ ms < k \leq m(s+1))$$

$$\leq (1 - q)^2$$

Similarly, $P(W_k \notin U,\ ms < k < m(s+i)) \leq (1 - q)^i$, which means that $P(W_k \in U$ f.o. (finitely often)$) = 0$. Hence $P(W_k \in U$ i.o.$) = 1$. Q.E.D.

4.15 Theorem: A state x is recurrent in the bilateral walk if and only if $\sum_{n=1}^{\infty} \mu^n * {}_x(\mu^n)(N(x)) = \infty$ for all neighborhoods N(x) of x.

Proof: The "==>" part is trivial by the Borel-Cantelli lemma. For the converse, suppose $\sum_n \mu^n * {}_x(\mu^n)(N(x)) = \infty$ for all N(x) of x. We claim that $x \in K$. If $x \notin K$, then for $y \in D$, there would exist a neighborhood $N_y(x)$ of x such that $\sum_n \mu^n(N_y(x)y^{-1}) < \infty$; this is because by Theorem (4.8), if $\sum \mu^n(N(x)y^{-1}) = \infty$ for all N(x), x, $y \in D$, then $x \in K$. Now there exist $N_y'(x)$, N(y) such that $N_y'(x)N(y)^{-1} \subset N_y(x)y^{-1}$ by Lemma (4.1). Since D is compact, there exist $N(y_i)$, $1 \leq i \leq n$, such that $D \subset \bigcup_{i=1}^n N(y_i)$. If N (x) (= a compact neighborhood of x) $\subset \bigcap_{i=1}^n N'_{y_i}(x)$, $\sum \mu^n(N_0(x)N(y_i)^{-1}) < \infty$, $1 \leq i \leq n$, so that $\sum \mu^n(N_0(x)D^{-1}) < \infty$. But

$$\infty = \sum_{n=1}^{\infty} \mu^n * {}_x(\mu^n)(N_0(x)) = \sum_{n=1}^{\infty} \int \mu^n(N_0(x)z^{-1})_x(\mu^n)(dz)$$

$$\leq \sum_{n=1}^{\infty} \mu^n(N_0(x)z_n^{-1}) < \infty.$$

where

$$\mu^n(N_0(x)z_n^{-1}) = \max_{z \in D} \mu^n(N_0(x)z^{-1}).$$

This maximum is attained since D is compact and the mapping $z \to \mu^n(N_0(x)z^{-1})$ is upper semicontinuous. (Since $N_0(x)$ is compact, $D - N_0 \equiv N_0^c$ is open so $\mu^n(N_0z^{-1}) = \mu^n((D - N_0^c)z^{-1}) = 1 - \mu^n(N_0^cz^{-1})$ and $-\mu^n(N_0^cz^{-1})$ is upper semicontinuous by Lemma (4.11)). The above contradiction proves that $x \in K$ and so x is recurrent in the bilateral walk. Q.E.D.

The following result states that all recurrent states in the three random walks are recurrent positive.

4.16 Theorem: Let $x \in K$. Then

(i) For all $N(x)$, $\overline{\lim}\, \mu^n(N(x)) > 0$

(ii) For all $N(x)$, $\overline{\lim}\, \mu^n(N(x)x^{-1}) > 0$

(iii) For all $N(x)$, $\overline{\lim}\, \mu^n *_x (\mu^n)(N(x)) > 0$.

Proof: (i) First, let U be any open set containing K = the kernel of D. We know (by [24, p. 141]) that $\mu^n(U) \to 1$ as $n \to \infty$. Now suppose for all $x \in K$, there is $N(x)$ such that $\overline{\lim}_{n\to\infty} \mu^n(N(x)) = 0$. Then since K is compact, $K \subset U = \cup_{i=1}^{k} N(x_i)$, for some k and $\overline{\lim}\, \mu^n(N(x_i)) = 0$; but this implies $\overline{\lim}\, \mu^n(U) = 0$, which is a contradiction. Hence there is $x \in K$ such that $\overline{\lim}\, \mu^n(N(x)) > 0$ for all $N(x)$. Next, let $I = \{x \in D;\ \overline{\lim}\, \mu^n(N(x)) > 0 \text{ for all } N(x)\}$. $I \neq \phi$ since $x \in I \cap K$. Let $z \in D$. Then, for all $N(xz)$, there exist $N(x)$, $N(z)$ such that $N(x)N(z) \subset N(xz)$. Now, choosing k so that $\mu^k(N(z)) > 0$, we have

$$\mu^{n+k}(N(xz)) \geq \int_{N(z)} \mu^n(N(xz)w^{-1})\mu^k(dw) \geq \mu^n(N(x))\mu^k(N(z)),$$

since for all $w \in N(z)$, $N(xz)w^{-1} \supset N(x)$; this means that $\overline{\lim}\, \mu^n(N(xz)) > 0$. Similarly, $\overline{\lim}\, \mu^n(N(zx)) > 0$. Hence I is an ideal and hence $K \subset I$.

(ii) We have $\mu^n *_x (\mu^n)(N(x)) = \int \mu^n(N(x)y^{-1})_x(\mu^n)(dy)$. Now, given $N(x)$, we can find $N_1(x)$ such that $N_1(x)x^{-1} \subset N(x)y^{-1}$ for all $y \in N_1(x)$; ([proof] for otherwise, there would exist a net $y_\alpha \to x$ such that $y_\alpha \in N_1(x)$ and $N_1(x)x^{-1} \cap (N(x)\, y_\alpha^{-1})^c \neq \phi$, so that we can find $z_\alpha \in N_1(x)x^{-1} \cap (N(x)y_\alpha^{-1})^c$ such that $z_\alpha x \in N_1(x)$,

$z_\alpha y_\alpha \notin N(x)$, which gives a contradiction (by using compactness of D)). Hence

$$\mu^n * {}_x(\mu^n)(N(x)) \geq \int_{N_1(x)} \mu^n(N_1(x)s^{-1})\, {}_x(\mu^n)(ds)$$

$$= \mu^n(N_1(x)x^{-1})\mu^n(x^{-1}N_1(x)),$$

which shows that $\overline{\lim}\ \mu^n * {}_x(\mu^n)(N(x)) > 0.$

Since from the structure of $K = E \times G \times F$ one easily checks that $N(x)x^{-1} \neq \phi$, (i) ==> (ii).

To obtain the number and structure of essential classes for the bilateral walk, we shall need the following theorem of Rosenblatt [24], Theorem 4.13, Ch. I.

4.17 <u>Theorem</u>: Suppose D is compact and let $K = E \times G \times F$, the kernel of D. (As always $D = \overline{\bigcup S_\mu^n}$, S_μ the support of μ). The sequence μ^n will not converge (in the weak topology) if and only if there is a closed normal subgroup $G' \subset G$, $G' \neq G$ such that $FE \subset G'$ and

$$(E \times G' \times F)S_\mu = E \times gG' \times F = E \times G'g \times F$$

where $g \notin G'$ and $G = \overline{\bigcup g^n G'}$

4.18 <u>Note</u>: Let η_e be the identity element of the group kernel of the compact abelian semigroup of measures $\overline{\{\mu^n : n \geq 1\}}$. Let S_{η_e} be the support of η_e. It can be shown (cf. proof of above theorem in [24]) that $S_{\eta_e} = E \times G' \times F$ where G' has all properties of the subgroup G' of the above theorem except possibly that of being a proper subgroup. If μ^n does not converge, then also $G' \neq G$.

Let K_2 be the kernel of $\overline{\bigcup S_\mu^{2n}}$. Then $K_2 = S_\beta$, the support of β, where $\beta = \lim \frac{1}{n} \sum_{j=1}^n \mu^{2j}$. Since β is idempotent, $K_2 \subset K$.

Since η_e and μ^2 commute, $\eta_e * \beta = \beta * \eta_e$. It follows that $S_\beta = E \times G_1 \times F$, $FE \subset G_1$, G_1 a closed subgroup of G. Clearly if μ^{2n} converges, then $\mu^{2n} \to \beta$.

Since the support of $\mu * \eta_e$ is $S_\mu(E \times G' \times F) = E \times gG' \times F$ and $S_{\beta * \eta_e} = (E \times \overline{\bigcup g^{2n}G'} \times F)$ and η_e is a cluster point of $\{\mu^n\}$, it can be easily shown that $(E \times \overline{\bigcup g^{2n}G'} \times F) \subset K_2$.

4.19 **Proposition:** Let K_2 be the kernel of $\overline{\bigcup S_\mu^{2n}}$, where S_μ is the support of any probability measure μ such that $D = \overline{\bigcup S_\mu^n}$, a compact semigroup. Then the number of sets of the form $\overline{\bigcup_n (S_\mu K_2)^n \, x \, (S_\mu K_2)^n} \equiv \overline{\bigcup S_\mu^n K_2 \, x \, K_2 S_\mu^n}$, for $x \in K$, is at most two, namely K_2 and $S_\mu K_2$ which are disjoint or equal and $K_2 \cup S_\mu K_2 = K$. The above sets are the essential classes for the bilateral walk induced on K by $\mu * \beta$ where $\beta = \lim_n \frac{1}{n} \sum_{j=1}^n \mu^{2j}$.

Proof: Clearly $K_2 = S_\beta \subset K$, where $\beta = \lim \frac{1}{n} \sum_{j=1}^n \mu^{2j}$. By (4.18), $K_2 = E \times G_1 \times F$, $FE \subset G_1$. We note that $K_2 S_\mu = S_\mu K_2$, $S_\mu^2 K_2 = K_2$ and $\overline{\bigcup (S_\mu K_2)^n} = S_\mu \cup S_\mu K_2 = K$. If $x \in K_2$, then $\overline{\bigcup (S_\mu K_2)^n \, x \, (S_\mu K_2)^n} = \overline{\bigcup S_\mu^n K_2 x \; K_2 S_\mu^n} = K_2$. [Note that K_2 and $S_\mu K_2$ are disjoint or equal; for if $f \in S_\mu$ and $fk \in K_2$, then $S_\mu fk \subset S_\mu^2 k \subset K_2$ and $S_\mu K_2 = S_\mu K_2 fk K_2 \subset K_2$]. Next let $x = (x_1, g_1, y_1)$, $y = (x_2, g_2, y_2) \in K - K_2 = S_\mu K_2$ so that $g_1, g_2 \notin G_1$. Then $z = (x_2, (y_1 x_2)^{-1} g_2^{-1}, y_2) \in K - K_2$. But $xz = (x_1, g_1 g_2^{-1}, y_2) \in S_\mu K_2 S_\mu K_2 = K_2$ and $g_1 g_2^{-1} \in G_1$. H follows that $G_1 g_1 G_1 = G_1 g_2 G_1$ and $\overline{\bigcup S_\mu^n K_2 x \; K_2 S_\mu^n} = \overline{\bigcup S_\mu^n K_2 y K_2 S_\mu^n} = \overline{\bigcup S_\mu^n K_2 (S_\mu K_2) K_2 S_\mu^n} = S_\mu K_2$.

4.20 Theorem: The essential classes of the bilateral walk W_n are at most two. They are of the form $S_\beta = K_2$ (= the kernel of $\overline{\cup s_\mu^{2n}}$) = $(E \times \overline{\cup g^{2n}G'} \times F)$ and $S_\beta S_\mu = K_2 S_\mu = (E \times g \overline{\cup g^{2n}G'} \times F) \subset \overline{\bigcup_{n=1} s_\mu^{2n+1}}$. In case both μ^n and μ^{2n} converge, then they become one class, $K_2 = K_2 S_\mu = K$.

Proof: Let us call $H \equiv \overline{\cup g^{2n}G'}$. By Proposition (4.19), $K_2 \cup K_2 S_\mu = K$ and by (4.18), $E \times \overline{\cup g^{2n}G'} \times F \subset K_2$. Also $S_\mu(E \times H \times F) =$ $S_\mu(E \times G' \times F)(E \times H \times F) = (E \times gG' \times F)(E \times H \times F) = (E \times gH \times F) \subset S_\mu K_2$. But since $(E \times H \times F) \cup E \times gH \times F) = K$, we have $K_2 = E \times H \times F$ and $S_\mu K_2 = E \times gH \times F$.

(Note that H and G are groups and $G = H \cup gH$ by Theorem (4.17).

Next for $x \in K_2$, $\overline{\cup s_\mu^n x s_\mu^n}$, an essential class for W_n, is contained in $\overline{\cup s_\mu^n K_2 s_\mu^n} = \overline{\cup s_\mu^{2n} K_2} = K_2$ and for $x \in S_\mu K_2$, $\overline{\cup s_\mu^n x s_\mu^n} \subset S_\mu K_2$.

Also for $x \in K_2$, $\overline{\bigcup_n (E \times gG' \times F)^n x (E \times gG' \times F)^n} =$

$\overline{\bigcup_n (E \times g^n G' \times F) x (E \times g^n e' \times F)} = E \times \overline{\bigcup_n g^n h g^n G'} \times F$, for some $h \in H$.

There is a net $g^{2k_\alpha} g'_\alpha \to h$, where $g'_\alpha \in G'$. But $\overline{\bigcup_n g^n g^{2k_\alpha} g'_\alpha g^n G'} =$

$\overline{\bigcup_n g^{2n+2k_\alpha} G'}$ being an ideal of $H = \overline{\cup g^{2n} G'}$, equals to H for every α.

Since $\cup g^n g^{2k_\alpha} g'_\alpha g^n G' \subset H$, we have $\overline{\cup g^n h g^n G'} \subset H$. Now if $w \in H$ and $w \notin \overline{\cup g^n h g^n G'}$, there must be a compact neighborhood $N(w)$ such that $N(w) \cap \overline{\cup g^n h g^n G'} = \phi$. But for each α, there is $g^{n_\alpha} g^{2k_\alpha} g'_\alpha g^{n_\alpha} g'' \in N(w)$ and hence an element of the form $g_1 h g_1 g' \in N(w)$ where $g^{n_\alpha} \to g_1$ and $g''_\alpha \to g'$, and this is a contradiction.

Hence for every $x \in K_2$, $\overline{\bigcup_n (E \times g^nG' \times F)x(F \times g^nG' \times F)} = E\times H\times F$

and for every $x \in S_\mu K_2 = E \times gH \times F$, $\overline{\bigcup_n (E \times g^nG' \times F)x(E \times g^nG' \times F)}$

$= E \times gH \times F$. So the bilateral walk induced on the kernel K by the measure $\mu * \eta_e$ (cf. Theorem (4.17)) has at most two essential classes. Since η_e is a cluster point of $\{\mu^n\}$, the essential classes of the bilateral walks induced by μ and $\mu*\eta_e$ coincide.

If S is compact abelian, then the kernel K becomes a compact group and the set of unconditionally recurrent states $R_u^r = R_u$ coincides with K as the following theorem shows. (In the abelian case we drop the superscripts indicating right or left and we write $R_u^r \equiv R_u^\ell = R_u$ and $R^r \equiv R^\ell = R$.)

4.21 __Theorem:__ Let S be compact abelian. Then $R_u = R = K = R_x$ for every $x \in D$, where $R_x = \{y \in D : P_x(Z_n \in N(y) \text{ i.o.}) = 1 \text{ for all } N(y) \text{ of } y\}$.

__Proof:__ Since S is abelian, K is a compact topological group. For any (relatively open) neighborhood $N \subset D$, by restricting our attention to D replacing S by D,

$$P(Z_n \in N \text{ i.o.}) = 0 \text{ or } 1, \tag{6}$$

since $P(z_n \in D^c$ for some $n) = 0$. (The proof of (6) (known as Hewitt-Savage 0-1 law) can be carried over to locally compact abelian semigroups from [4, p. 255]). Next assume $R_u = \phi$. Then for each $x \in K$, there is a neighborhood N_x of x such that $P(Z_n \in N_x \text{ i.o.}) < 1$ and hence $P(Z_n \in N_x \text{ i.o.}) = 0$ by (6). Since K is compact, there is a finite cover of K consisting of these N_x's. Say, $K \subset U = \cup_{i=1}^n N_x^{(i)}$.

Then $P(Z_n \in U \text{ i.o.}) = 0$, which contradicts Prop. (4.9). Hence $R_u \cap K \neq \phi$.

By the Borel-Cantelli lemma and Theorem (4.5), $R_u \subset K$ and since

R_u is easily shown to be an ideal, $R_u = K$. [Note that $P(Z_n \in N(xy) \text{ i.o.}) \geq P(Z_n \in N(x)N(y) \text{ i.o.}) \geq P(Z_k \in N_x)P(Z_k^{-1}Z_n \in N_y \text{ i.o.})$ if $N(x)N(y) \subset N(xy)$].

Next, we observe that for $k \in K$, $x \in D$, $x^{-1}N(k) \neq \phi$. For if e is the identity of K, then $ex \in K(N(k) \cap K) = K$ so that $ex = k_1^{-1}n_k$ for some $k_1 \in K$ and $n_k \in N(k) \cap K$, so that $k_1ex = k_1x = n_k \in N(k)$. Hence by Theorem (4.16), $\overline{\lim}\ \mu^n(x^{-1}N(k)) > 0$ and using (6), $P_x(Z_n \in N(k) \text{ i.o.}) = 1$.

4.22 Note: Theorem (4.20) describing the essential classes for the bilateral walk W_n becomes simplified in the compact abelian case. Since $W_n = X_\circ Z_{2n}$ in this case. Then K_2, the kernel of $\overline{\bigcup S_\mu^{2n}}$, is a subgroup of K of index two and $K = K_2 \cup S_\mu K_2$. Also $K = \overline{\bigcup(S_\mu K_2)^n}$, $K_2 = \overline{\bigcup(S_\mu K_2)^{2n}}$ and $S_\mu K_2 = \overline{\bigcup(S_\mu K_2)^{2n+1}} = R_u^b$. (For the last equality concerning the set of unconditional recurrent states for W_n, see the related result for abelian groups in the next Section, Corollary (5.6)).

5. Miscellaneous results.

In this Section we will give some general results concerning the recurrence concepts and we will prove the equivalence of unilateral and bilateral recurrence in the case of abelian groups.

5.1 Recurrent Points and Points of Sure Return:

The following results (due to Rosenblatt [28]) are actually valid for general Markov transition functions $P(x,\cdot)$ and general locally compact 2nd countable spaces. Let us recall that a point $x \in D$ is said to be a point of sure return if for each neighborhood $N(x)$ of x

$$P_x(Z_n \in N(x) \text{ for some } n \geq 1) = 1$$

In the previous sections of this chapter we established the equivalence of being conservative and recurrent. The result of Rosenblatt establishes the equivalence of being recurrent and a state of sure return.

Let A be a Borel set. Then the probability of hitting A for the first time on step $k(\geq 1)$ given that one starts from x at time 0 is

$$(P1_{A^c})^{k-1}P1_A(x) = (P1_{A^c})^{k-1}P(1 - 1_{A^c})(x)$$

$$= (P1_{A^c})^{k-1}(x) - (P1_{A^c})^k(x), \text{ where}$$

$P1_A(x) = P(x,A)$ stands for the transition function of any of the three random walks and the operator $P1_A$ is defined by $P1_Af \equiv \int_A f(y)P(\cdot,dy)$. The probability of never hitting A given that one starts from x is

$$\lim_{k\to\infty} (P1_{A^c})^k(x) = \lim_{k\to\infty} \int_{A^c} P(x,dz_1)\int_{A^c}\cdots\int_{A^c} P(z_{k-2},dz_{k-1})P(z_{k-1},A^c).$$

Thus x is a point of sure return if for each neighborhood N_x of x

$$P(Z_k \in N_x \text{ for some } k \geq 1|X_0 = x) = 1 - \lim_{m\to\infty}(P1_{N_x^c})^m(x) = 1$$

5.2 Lemma: Let x be a point of sure return. Then if for some integer $j \geq 1$

$$P(Z_k \in N_x \text{ for } j \text{ distinct } k\text{'s} \geq 1|X_0 = x) = 1,$$

for each neighborhood N_x of x, it follows that for each neighborhood N_x of x

$$P(Z_k \in N_x \text{ for } j + 1 \text{ distinct } k\text{'s} \geq 1|X_0 = x) = 1.$$

Proof: Let

$$P_B(y,A) = P(y,A) + \int_{B^c} P(y,dy_1)P(y_1,A)$$

$$+ \int_{B^c} P(y,dy_1) \int_{B^c} P(y_1,dy_2)P(y_2,A)+\cdots$$

for $A,B \in \mathcal{B}$. If $A \subset B$, this is the probability of first hitting the set B (at some time $k \geq 1$) in the set A given that one starts at x.

If $A \subset B^c$, $P_B(y,A)$ can be interpreted as the mean number of hits of A before hitting B. The assumption that x is a point of sure return means that

$$P_{N_x}(x,N_x) = 1,$$

for each neighborhood N_x of x. Suppose that $P_{N_x}(x, \{x\}) = q \geq 0$

where $\{x\}$ is the set containing the one point x. The assumption that j distinct returns to N_x are sure implies that

$$P(Z_{n_1} = x, Z_{n_\alpha} \in N_x, \alpha = 2, \ldots, j+1 \mid X_0 = x) = q,$$

where $1 \leq n_1 < n_2 < \ldots < n_{j+1}$ are the first $j + 1$ distinct returns to N_x. Also

$$P_N(x, N_x - \{x\}) = 1 - q .$$

Let $N_x(d)$ be the neighborhood of x consisting of the points at a distance from x less than d. (Here is the first time in this Chapter where metrizability of S was used). Then

$$P_{N_x}(x, N_x - N_x(d)) \uparrow 1 - q,$$

as $d \downarrow 0$. The assumption that j distinct returns to $N_x(d)$ from x are sure implies that

$$P(Z_k \in N_x(d) \text{ for } j \text{ distinct } k\text{'s} \geq 1 \mid X_0 = y) = 1,$$

for almost all y in $N_x - N_x(d)$ with respect to $P_{N_x}(x,\cdot)$. But then if $n(\geq 1)$ is the first hitting time of N_x

$$P_{N_x}(x, N_x - N_x(d)) = P(Z_\ell \notin N_x, 1 \leq \ell < n, Z_n \in N_x - N_x(d), Z_k \in N_x(d)$$

$$\text{for } j \text{ distinct } k\text{'s} \geq n \mid X_0 = x) \uparrow 1 - q ,$$

as $d \downarrow 0$. This yields the desired conclusion, that $j + 1$ distinct returns to N_x are sure given that one starts at x.

5.3 Theorem (Rosenblatt). Let S be a locally compact 2nd countable semigroup. Then in all the three random walks on S, x is a point of sure return if and only if it is recurrent.

Lemma (5.2) implies that if x is a point of sure return, then by induction it is a point of j distinct sure returns for every positive integer j and for each neighborhood of x. Thus x is a recurrent point.

Let $P^{(n)}(x,\cdot) = P^n(x,\cdot)$ be the nth transition function for any of the three random walks on S. Let

$$(7)\qquad P_B(y,A) = P(y,A) + \int_{B^c} P(y,dy_1)P(y_1,A) + \int_{B^c} P(y,dy_1) \int_{B^c} P(y_1,dy_2)P(y_2,A) + \ldots$$

for A, B ε $\mathcal{B}$ (= the Borel σ-field of D). If A $\subset$ B, this is the probability of first hitting the set B (at some time $k \geq 1$) in the set A given that one starts at x. Consider the measure

$$Q(x,\cdot) = \sum_{n=1}^{\infty} \frac{1}{2^n} P^n(x,\cdot)$$

whose support is $\overline{\bigcup S_\mu^n x} = Dx$ whenever D is a semigroup and $P(x,\cdot)$ is induced by μ of support S_μ.

5.4 Lemma: For any x and B ε $\mathcal{B}$, the measures $P_B(x,\cdot)$ and $Q(x,\cdot)$ are equivalent (that is, mutually absolutely continuous) as measures on B.

Proof: If $P_B(x,A) > 0$ for some A $\subset$ B then one of the terms on the right of (7) must be positive, say the k-th. But the k-th term is less than or equal to $P^{(k)}(x,A)$. Thus $P_B(x,A) > 0$ implies that $Q(x,A) > 0$ and so $P_B(x,\cdot)$ is absolutely continuous with respect to $Q(x,\cdot)$. Conversely if $Q(x,A) > 0$ for A $\subset$ B, then $P^{(k)}(x,A) > 0$ for some k and hence $P_B(x,A) > 0$. Thus $Q(x,\cdot)$ is absolutely continuous with respect to $P_B(x,\cdot)$ on B.

The following Lemma is also of some interest.

5.5 Lemma: Let x be a point of sure return (and so recurrent). Consider a point $y \neq x$ with N_x and N_y disjoint neighborhoods of x and y respectively. Then

$$P_{N_x}(z,N_x) = 1,$$

for almost all $z \in N_y$ with respect to $P_{N_y}(x,\cdot)$.

Proof: Notice that

$$P(Z_j \in N_x \text{ for some } j \geq 1 | X_0 = x)$$

$$\leq P(Z_j \in N_y,\ N_k \in N_x \text{ for some } j,\ k \text{ with } 1 \leq j < k | X_0 = x)$$

$$+ P(Z_j \notin N_y \text{ for all } j \geq 1 | X_0 = x),$$

since x is recurrent. This implies that

$$P_{N_x}(x,N_x) \leq \int_{N_y} P_{N_y}(x,dz) P_{N_x}(z,N_x) + (1 - P_{N_y}(x,N_y)).$$

However, since $P_{N_x}(x,N_x) = 1$,

$$P_{N_y}(x,N_y) \leq \int_{N_y} P_{N_y}(x,dz) P_{N_x}(z,N_x) \leq P_{N_y}(x,N_y),$$

and the conclusion follows. Notice that a simple modification of the argument given here implies that under the assumptions of the proposition

$$P(Z_j \in N_x \text{ infinitely often} | X_0 = z) = 1,$$

for almost all $z \in N_y$ with respect to $P_{N_y}(x,\cdot)$.

The following two interesting results are actually valid for general Markov transition functions $P(x,\cdot)$ such that $Pf(x) = \int P(x,dy)f(y)$ is a bounded continuous function for every bounded continuous f on S, S being a locally compact 2nd countable space (not necessarily a semigroup!).

5.6 Proposition: (Rosenblatt). Let x be a recurrent point and N a neighborhood not containing x. Then almost all $z \in N$ with respect to $P_N(x,\cdot)$ are points of sure return and hence recurrent.

Proof; Assume that $P(x,\cdot)$ is not trivial, that is, $P(x,\{x\}) < 1$ since otherwise the result is obvious. The proof is indirect. If the conclusion is false, there is a subset M of N of positive $P_N(x,\cdot)$ measure consisting entirely of points that are not of sure return. Let z be any point of this set. Then

$$P(Z_j \notin N_z \text{ for all } j \geq 1 | X_0 = z) > 0,$$

for all sufficiently small neighborhoods N_z of z. Thus for each neighborhood N_x of x, there is a neighborhood N_z of z such that

$$P(Z_j \in N_x \text{ infinitely often, } Z_j \notin N_z \text{ for all } j \geq 1 | X_0 = z)$$

$$= F(z,N_z) > 0$$

for almost all such z. If

$$F_n(z,N_z) = \int_{N_x} P_{N_z \cup N_x}(z,du_1) \int_{N_x} P_{N_z \cup N_x}(u_1,du_2) \ldots P_{N_z \cup N_x}(u_{n-1},N_x),$$

then

$$F(z,N_z) = \lim_{n\to\infty} F_n(z,N_z).$$

We can find a sequence of disjoint neighborhoods $N_1,\ldots,N_{j-1}$ not containing x or z such that for some sufficiently small neighborhood N_z of z

$$\int_{N_1} P(x,du_1) \int_{N_2} P(u_1,du_2) \ldots P(u_{j-1},N_z) > \delta > 0.$$

Since P takes continuous functions into continuous functions, one
can choose a neighborhood of x, N_x, sufficiently small so that it
is disjoint from $\bigcup_{\alpha=1}^{j-1} N_\alpha$ and N_z such that for all $y \in N_x$

$$\int_{N_1} P(y,du_1) \int_{N_2} P(u_1,du_2)\ldots P(u_{j-1},N_z) > \frac{\delta}{2} > 0.$$

Thus,

$$P_{N_z \cup N_x}(y,N_z) > \frac{\delta}{2}, \qquad y \in N_x$$

so that

$$P_{N_z \cup N_x}(y,N_x) < 1 - \frac{\delta}{2} = \alpha < 1, \ y \in N_x.$$

But then

$$F_n(z,N_z) < \alpha \ F_{n-1}(z,N_z) \qquad n = 2,3,\ldots$$

The transitions functions of the three random walks induced
by μ, clearly map continuous bounded functions into continuous
bounded functions. If this condition is not satisfied by the Markov
transition $P(x,\cdot)$, Proposition (5.6) is no longer true as the following example of Rosenblatt shows.

Example: Let the states of a Markov process be the positive integers
$1,2,\ldots$ and ∞. The topology on the finite integers is discrete and
one has the one point compactification at ∞ (neighborhoods of ∞ of the
form $\{n,n+1,\ldots,\infty\}$ with n a finite integer). Let the transition
probabilities be

$$P_{n,n-1} = q, \qquad P_{n,n+1} = p$$

$$\text{with } p + q = 1, \qquad \tfrac{1}{2} < p < 1, \qquad n = 2,3,\ldots$$

$$P_{1,1} = q, \qquad P_{1,2} = p$$

$$P_{\infty,\infty} = p, \qquad P_{\infty,0} = q.$$

The point ∞ is then a recurrent point in our sense. However, the finite integers it leads to are all non-recurrent. Of course, the transition function does not take continuous functions into continuous functions.

5.7 Theorem: (Rosenblatt). Let the transition function P map bounded continuous functions into bounded continuous functions on a locally compact 2nd countable space D. (D not necessarily a semigroup). Assume that x is a recurrent point. Then almost all z with respect to Q(x,) are recurrent.

This is an immediate consequence of Proposition (5.6) and Lemma (5.5). It is quite interesting since it indicates that some aspects of what is true for recurrent states of a countable state Markov chain still holds for general recurrent states. This is that almost all states y that can be reached from x (in the sense that y is in the measure theoretic support of $Q(x,\cdot)$ are recurrent if x is recurrent.

Recurrence in abelian groups.

In order to prove that recurrence in the unilateral walk implies recurrence in the bilateral for the abelian group case, we shall need the following theorem which is of interest by itself.

We observe that in the abelian case the bilateral walk $W_n = X_{-n}..X_{-1}X_0X_1..X_n$ reduces to the products $W_n = X_0Z_{2n}$, where $Z_{2n} = X_1X_2...X_{2n}$.

5.8 Theorem: Let S be an abelian group and assume that the unilateral walk is recurrent. Then $\overline{\bigcup_{n-1}^{\infty} S_\mu^{2n}}$ is a subgroup of D and $\sum_{n=1}^{\infty}\mu^{2n}(N(e)= \infty$ for every neighborhood N(e) of the identity e in D. (S_μ = Support (μ)).

Moreover,

$$P_x(Z_{2n} \in N(x) \text{ i.o.}) = 1 \text{ for every } x \in D \text{ and every } N(x) \text{ of } x.$$

Proof: (By contradiction). Assume there is $N_o(e)$ such that $\sum \mu^{2n}(N_o) < \infty$. Let N be any neighborhood of e. Then also

$$(8) \qquad \sum \mu^{2n}(N \cap N_o) < \infty$$

Clearly then

$$(9) \qquad \sum \mu^{2n+1}(N) = \infty, \text{ (for every neighborhood N of e).}$$

Let k be an arbitrary positive integer. Since (9) holds we have

$\sum_{j=1}^{k} \sum_{i=o}^{\infty} P(Z_{2j-1\,+\,2ik} \in N) = \infty$. We can find an integer m, $1 \leq m \leq k$, such that

$\sum_{i=o}^{\infty} P(Z_{2m-1\,+\,2ik} \in N) = \infty$. Now

$$1 \geq P(Z_{2n+1} \in N \text{ finitely often})$$

$$\geq \sum_{i=o}^{\infty} P(Z_{2m-1\,+\,2ik} \in N,\ Z_{2n+1} \notin N \text{ for all } 2n+1 \geq 2m + 2(i+1)k-1)$$

$$= \sum P(Z_{2m-1+2ik} \in N,\ Z^{-1}_{2m-1+2ik} Z_{2n+1} \notin N^{-1}N \text{ for all } 2n+1 \geq 2m+2(i+1)k-1)$$

$$\geq \sum P(Z_{2m-1+2ik} \in N,\ X_{2m+2ik} \cdots X_{2n+1} \notin N^{-1}N \text{ for all } 2n+1 \geq 2m+2(i+1)k-1) \quad \text{(shifting by } 2m+2ik-1)$$

$$\geq \quad P(X_1 X_2 \ldots X_{2k'} \notin N^{-1}N \text{ for all } k' \geq k) \sum_{i=o} P(Z_{2m-1+2ik} \in N)$$

which means that $P(Z_{2k'} \notin N^{-1}N \text{ for all } k' \geq k) = 0$. Since we can find an N_e such that $N_e^{-1}N_e \subset N$, we obtain

$$P(Z_{2k'} \in N \text{ finitely often}) \leq P(Z_{2k'} \in N^{-1}N \text{ finitely often}) = 0$$

which implies that $\sum \mu^{2n}(N) = \infty$. (See Theorems (4.5) and (4.8) for a similar proof and particular details).

From Theorem (5.8) the following summary for the locally compact abelian group case (resp. the compact abelian semigroup case) follows.

5.8A Corollary:

(a). Let S be a locally compact abelian group. Then either one of R_u^r, R_u^b, R^r, R^b is non-empty in which case $R_u^r = R^r = R^b = D =$ a group, and $R_u^b = \overline{\cup S_\mu^{2n+1}}$ and the essential classes for $W_n = X_0 Z_{2n}$, ($Z_{2n} = X_1 X_2 \ldots X_{2n}$), are $\overline{\cup S_\mu^{2n}}$ = a subgroup of D, and $\overline{\cup S_\mu^{2n+1}}$, or all walks on S are non-recurrent (all above classes are empty).

(b). The recurrence case holds for compact abelian semigroup S with the role of D above replaced by K = the group kernel of D, the role of S_μ by $K_2 S_\mu$ (K_2 = the kernel of $\overline{\cup S_\mu^{2n}}$, $K_2 \subset K$), with the at most two essential classes for $W_n = X_0 Z_{2n}$ being $\overline{\cup (S_\mu K_2)^{2n}} \equiv K_2$ and $\overline{\cup (S_\mu K_2)^{2n+1}} \equiv S_\mu K_2$, $K_2 \cup S_\mu K_2 = K$.

Proof: (a) It is clear that $R_b \subset \overline{\cup S_\mu^{2n+1}}$. Conversely, let $x \in \overline{\cup S_\mu^{2n+1}}$. Then

$$P(X_o Z_{2n} \in N(x) \text{ i.o.}) \geq \int_{N(x)} P_x(Z_{2n} \in N(x) \text{ i.o.})\, \mu^{2k+1}(dx), \quad \text{for}$$

some odd 2k+1, and so the probability on the left equals 1 (zero-one-law). (cf. Th. (5.8)). Next, we only verify the claims about the essential classes for W_n. Clearly $\overline{\cup S_\mu^n x S_\mu^n} = \overline{\cup S_\mu^{2n} x}$ and every element in $\overline{\cup S_\mu^{2n} x} - \cup S_\mu^{2n} x$ leads to x, so that the essential class $C(x) \equiv \overline{\cup S_\mu^{2n} x}$. Since $\overline{\cup S_\mu^{2n}}$ is a group (Th. (5.8)), every two elements $f_1, f_2 \in S_\mu$ communicate in the bilateral walk since $f_2 \in S_\mu^{-1} f_1 S_\mu \subset \overline{\cup S_\mu^{-2n} S_\mu^{2n} f_1} = \overline{\cup S_\mu^{-2n} f_1}$. Hence every element in $\overline{\cup S_\mu^{2n+1}}$ gives rise to the same coset of $\overline{\cup S_\mu^{2n}}$.

Since $\overline{\bigcup S_\mu^{2n}}\, R_b = \overline{\bigcup S_\mu^{2n}}\ \overline{\bigcup S_\mu^{2n+1}} \subset \overline{\bigcup S_\mu^{2n+1}} = R_u^b$, we have $R_u^b = D$ if and only if e (= identity) $\varepsilon\ R_u^b$. In the simple example of the two element group {-1,1} under multiplication with $S_\mu = \{-1\}$ the even powers of S_μ do not generate D.

(b). In Section 4 we have shown the equivalence of the recurrence concepts in the unilateral and bilateral random walks for compact semi-groups, where the recurrent states of the three random walks form precisely the kernel K of D. It was shown that both walks are recurrent and that x is essential (for W_n) if and only if x is x-recurrent if and only if $x \,\varepsilon\, K$ if and only if $\sum P_b(x, N_x) = \infty$ for every neighborhood N_x of x. Since the identity e in K is recurrent for W_n, $\sum \mu^{2n}(N(e) = \infty$ for all N(e) and hence $e \,\varepsilon\, K_2$. It follows that $K_2 = K \cap \overline{\bigcup S_\mu^{2n}}$. The essential classes for W_n are $\overline{\bigcup S_\mu^n x S_\mu^n} = \overline{\bigcup S_\mu^{2n} x} = K_2$ for $x \,\varepsilon\, K_2$ and $\overline{\bigcup S_\mu^{2n}}\ ex = K_2 x$ for $x \notin K_2$. In the group K consider the random walk induced by the measure whose support is $S_\mu K_2$. (K_2 supports $\lim\, (1/n) \sum_{j=1}^{n} \mu^{2j}$, [24], Chapter V). The even powers of $S_\mu K_2$ generate K_2 and all powers of $S_\mu K_2$ generate K. By the group case, the index of K_2 is 2 and the essential classes of the new walk are identical with those of the walk induced by S_μ. In fact the essential classes are K_2 and $K_2 S_\mu$ since $\overline{\bigcup S_\mu^n K_2 f_1^{2k} f_2 K_2 S_\mu^n} \subset \overline{\bigcup S_\mu^n K_2 f_2 K_2 S_\mu^n} \subset \overline{\bigcup S_\mu^{2n} K_2 S_\mu} = K_2 S_\mu$, where $f_1^{2k} f_2 \,\varepsilon\, S_\mu^{2k} S_\mu$, and any two elements in the support $K_2 S_\mu$ communicate (cf. proof of group case). For the locally compact abelian group case, $K_2 \equiv \overline{\bigcup S_\mu^{2n}}$ so that the essential classes are $\overline{\bigcup S_\mu^{2n}}$ and $\overline{\bigcup S_\mu^{2n+1}}$ (which must be disjoint or equal if S_μ is to generate a recurrent walk on D).

Note: From Theorem (5.8) it follows that for arbitrary locally compact group, if a unilateral walk Z_n is recurrent (i.e. $\{x \in D;\ x \in R^r\} = \emptyset$), then the walk generated by μ^2, $(X_{-0}X_0)Z_{2n}$, is also recurrent (and conversely), and $\overline{\bigcup S_\mu^{2n}}$ is a subgroup of the group D. Hence $\overline{\bigcup S_\mu^{2n} S_\mu} = \overline{\bigcup S_\mu^{2n+1}}$ and $\overline{\bigcup S_\mu^{2n}}$ are disjoint or equal (by properties of cosets) and their union is D. (The cosets of $\overline{\bigcup S_\mu^{2n}}$ are the essential classes for the μ^2-walk; hence a necessary condition that a locally compact group admit a unilateral recurrent walk is that $\overline{\bigcup S_\mu^{2n}}$ and $\overline{\bigcup S_\mu^{2n+1}}$ must be disjoint or equal. If in addition the group is abelian, then the bilateral walk W_n and the μ^2-walk have the same transition functions (and the same essential classes) and hence recurrence in the bilateral walk is equivalent to recurrence in the unilateral. In the recurrent case, $\overline{\bigcup S_\mu^{2n+1}}$ is a single coset of $\overline{\bigcup S_\mu^{2n}}$ since any two elements $f_1, f_2 \in S_\mu$ communicate in the $W_n = X_0 Z_{2n}$ (and in the μ^2-walk) and the essential classes for $W_n = X_0 Z_{2n}$ (where $Z_{2n} \equiv X_1 X_2 \ldots X_{2n}$), and the μ^2-walk $(X_{-0}X_0)Z_{2n}$, are at most two.

A generalization of the compact case: Conditions (CR) and (CL). Every compact semigroup satisfies conditions (CR) and (CL). (cf. (2. 6)). The following result is an interesting generalization of the compact semigroup case and taken from [21].

5.9 Theorem: Suppose that the semigroup S has (CL) and (CR). Suppose there exists $x \in S$ with the property:

(*) $\sum_{n=1}^{\infty} \mu^n(N(x)) = \infty$ for every open $N(x)$ containing x. Then S has a completely simple kernel K which consists of only and all those points of S with property (*).

Proof. Let I be the set of all points of S with property (*). Then I is an ideal of S. To see this, let $z \in S$ and $y \in I$. Then given any open V containing zy, there exist open sets N(z) and N(y) containing z and y respectively such that $N(z)N(y) \subset V$. Let k be a positive integer such that $\mu^k(N(z)) > 0$. Then

$$\sum_{n=1}^{\infty} \mu^{n+k}(V) = \sum_{n=1}^{\infty} \int \mu^n(w^{-1}V)\, \mu^k(dw) = \infty$$

since for $w \in N(z)$, $N(y) \subset w^{-1}V$. Hence $zy \in I$. Similarly, $yz \in I$. Thus I is an ideal of S.

As usual, let $X_1, X_2, \ldots$ be a sequence of independent identically distributed (with distribution μ) random variables with values in S and let $Z_n = X_1X_2\ldots X_n$. Then given any positive integer k, we can find a positive integer m, $1 \leq m \leq k$ such that

$$\sum_{i=0}^{\infty} P(Z_{m+ik} \in N(x)) = \sum_{i=0}^{\infty} \mu^{m+ik}(N(x)) = \infty.$$

Now, we have

$$1 \geq P(Z_n \in N(x) \text{ finitely often})$$

$$\geq \sum_{i=0}^{\infty} P(Z_{m+ik} \in N(x),\ Z_n \notin N(x) \text{ for all } n \geq m+(i+1)k)$$

$$\geq \sum_{i=0}^{\infty} P(Z_{m+ik} \in N(x),\ X_{m+ik+1}\ldots X_n \notin N(x)^{-1}N(x) \text{ for all } n \geq m+(i+1)k)$$

$$= P(Z_n \notin N(x)^{-1}N(x) \text{ for all } n \geq k).\ \sum_{i=0}^{\infty} P(Z_{m+ik} \in N(x)).$$

Hence for each positive integer k,

$$P(Z_n \notin N(x)^{-1}N(x) \text{ for all } n \geq k) = 0.$$

Therefore,

(10) $P(Z_n \in N(x)^{-1}N(x) \text{ infinitely often}) = 1$ for every open set N(x) containing x.

From (10) and (CL), it follows that $x \in xS$. We claim that xS is a minimal right ideal. To prove this, let $y \in xS$. Suppose $x \notin yS$. Then by (CL), there exist open sets V_y and V_x containing y and x respectively such that $V_y^{-1}V_x$ is empty. Let $y = xz$. Let W_x and W_z be open neighborhoods of x and z such that $W_x \subset V_x$ and $W_xW_z \subset V_y$. Then $W_z^{-1}[W_x^{-1}W_x]$ is also empty. Now there is a positive integer k such that $P(Z_k \in W_z) > 0$. But,

$$P(Z_k \in W_z) = P(Z_k \in W_z,\ Z_n \in W_x^{-1}W_x \quad \text{infinitely often})$$

$$\leq P(Z_k \in W_z,\ X_{k+1}\ldots X_n \in W_z^{-1}[W_x^{-1}W_x] \text{ infinitely often})$$

$= 0$, a contradiction. Hence $x \in yS$ for every $y \in xS$. This means that xS is a minimal right ideal. Similarly, Sx is a minimal left ideal. By [1], S has a kernel (which is the union of all minimal left ideals) which is completely simple. Since for each $y \in I$, $y \in yS \subset$ the kernel, $I = K$. Q.E.D.

REFERENCES

1. J.F. Berglund and K.H. Hofmann: Compact semitopological semigroups and weakly almost periodic functions. Lecture Notes in Math. no. 42, Springer-Verlag, New York, 1967.

2. A. Brunel and D. Revuz: Un critere probabiliste de compacite des groupes. Ann. of Probability, 2 (1974), 745-746.

2A. A. Brunel, P. Crepel, Y. Guivarc'h and M. Keane, Marches aleatoires recurrentes sur les groupes localement compacts. C. R. Acad. Sc. Paris, t 275(1972), 1359-1361.

3. K.L. Chung and W.J. Fuchs: On the distribution of values of sums of random variables. Mem. Amer. Math. Soc. No. 6 (1951). MR 12, 722.

4. K.L. Chung: A course in probability theory. Academic Press, 2nd Ed. New York, 1974.

5. A.H. Clifford and G.B. Preston: The algebraic theory of semigroups. I, II. Math. Surveys, no. 7, Amer. Math. Soc., Providence, R.I., 1961, 1967.

6. R.M. Dudley: Random walks on abelian groups. Proc. Amer. Math. Soc. 13 (1962), 447-450.

7. R.M. Dudley: Pathological topologies and random walks on abelian groups. Proc. Amer. Math. Soc. 15(1964) 231-238. MR 28, 5479.

8. U. Grenander: Probabilities on algebraic structures. Almqvist and Wiksell, Stockholm, 1963.

8A. Y. Guivarc'h and M. Keane, Transience des marches aleatoires sur les groupes nilpotents, Asterisque 4 (Soc. Math. de France, Paris, 1973.)

9. G. Högnäs: Marches aleatoires sur un demi-groupe compact. Ann. Inst. Henri Poincare, Section B, 10 (1974), 115-154.

10. H. Kesten and F. Spitzer: Random walks on countably infinite abelian groups. Acta Math. 114 (1965), 237 -265.

11. H. Kesten: The Martin boundary of recurrent random walks on countable groups. Proc. 5th Berkeley Symp. Math. Statist. Prob. 2 (1967), 51-74.

12. H. Kesten: Symmetric random walk on groups. Trans. Amer. Math. Soc. 92 (1959), 336-354.

13. R.J. Koch and A.D. Wallace: Maximal ideals in compact semigroups. Duke Math. J. 21 (1954), 681-685. MR, 16, 112.

14. J. Larisse: Marches au hasard sur les demi-groupes discrets. I,II. Ann. Inst. H. Poincaré 8(1972), 107-175.

15. J. Larisse: Marches au hasard sur les demi-groupes discrets, III. Ann. Inst. Henri Poincaré 8(1972), 229-240.

16. R. M. Loynes, Products of independent random elements in a topological group, Z. Wahrscheinlichkeitstheorie 1(1963). 446-455, MR 27,6293.

17. P. Martin-Löf: Probability theory on discrete semigroups, Z. Wahrscheinlichkeitstheorie 4(1966), 78-102. MR 32,1740.

18. A. Mukherjea: T. C. Sun and N. A. Tserpes, Random walks on compact semigroups, Proc. Amer. Math. Soc. 39(1973), 599-605.

19. A. Mukherjea and N. A. Tserpes: Bilateral random walks on compact semigroups, Proc. Amer. Math. Soc. 47(1975), 457-466.

20. A. Mukherjea and N. A. Tserpes: Some problems on random walks on semigroups, to appear in Proc. of the Caratheodory Symp. Athens.

21. A. Mukherjea: Limit theorems for probability measures on non-compact groups and semigroups, Z. Wahrscheinlichkeitstheorie 33, (1976), 273-284.

22. S. C. Port and C. J. Stone: Potential theory of random walks on abelian groups. Acta Math. 122(1969), 19-114.

23. S. C. Port and C. J. Stone: Infinitely divisible processes and their potential theory, I,II. Ann. Inst. Fourier, Grenoble - 21, 2 (1971), 157-275 and 21,4 (1971), 179-265 respectively.

23A. D. Revuz : Markov Chains. North-Holland Publishing Company, Amsterdam, 1975.

24. M. Rosenblatt: Markov Processes: Structure and asymptotic behavior. Springer-Verlag, New York, 1971.

25. M. Rosenblatt: Limits of convolution sequences of measures on a compact topological semigroup. J. Math. Mech. 9(1960), 293-306.

26. M. Rosenblatt: Stationary measures for random walks on semigroups. Proc. of a Symp. on Semigroups at Wayne State U.(K.W. Foley ed.) Academic Press, N.Y. 1969.

27. M. Rosenblatt: Invariant and subinvariant measures of transition probability functions acting on continuous functions. Z. Wahrscheinlichkeitstheorie 25(1973), 209-221.

28. M. Rosenblatt: Recurrent points and transition functions acting on continuous functions. Z. Wahrscheinlichkeitstheorie 30(1974), 173-183.

29. T. C. Sun, A. Mukherjea and N. A. Tserpes: On recurrent random walks on semigroups, Trans. Amer. Math. Soc. 185(1973), 213-228.

INDEX